ENVIRONMENTAL DESIGN RESEARCH DIRECTIONS

ENVIRONMENTAL DESIGN RESEARCH DIRECTIONS

Process and Prospects

Gary T. Moore,
D. Paul Tuttle,
and Sandra C. Howell

*Published in Cooperation with
the Environmental Design
Research Association, Inc.*

Environment, Behavior, and Design
Ervin H. Zube & Gary T. Moore
Consulting Editors

PRAEGER SPECIAL STUDIES • PRAEGER SCIENTIFIC

New York • Philadelphia • Eastbourne, UK
Toronto • Hong Kong • Tokyo • Sydney

Library of Congress Cataloging in Publication Data

Moore, Gary T.
  Environmental design research directions.

  "Co-published with Environmental Design Research
Association."
  Bibliography: p.
  Includes index.
  1. Environmental engineering—Research. I. Tuttle,
D. Paul. II. Howell, Sandra C. III. Environmental
Design Research Association. IV. Title.
TA170.M66 1984   620.8'072   84-18201
ISBN 0-03-000522-1

Published in 1985 by Praeger Publishers
CBS Educational and Professional Publishing
a Division of CBS Inc.
521 Fifth Avenue, New York, NY 10175 USA

56789 052 987654321

Printed in the United States of America
on acid-free paper

Dedicated to the memory of
Donald Appleyard,
Thomas Byerts,
and
Kevin Lynch

Contents

Preface and Acknowledgments

PURPOSE

The purpose of *Environmental Design Research Directions: Process and Prospects* is to provide a synthesis of the environment, behavior, and design field, to guide and inform future research and environmental application, and to present a set of research directions addressing these issues. The multidisciplinary field most identified with confronting environment, behavior, and design issues has been variously termed environment-behavior studies, environmental psychology, or environmental design research—the study of the mutual relations between the physical environment and human behavior and the application of knowledge thus gained to improving the quality of life through better informed environmental policy, planning, design, and education.

The book attempts to articulate environmental design research directions most needing immediate and sustained attention in order to help further structure the field, to aid in focusing research activities and resources, to encourage others interested in doing research or in applying its results, and to help structure education.

This monograph builds on a previous report, *Environmental Design/Research* by Marguerite Villecco and Michael Brill, published by the National Endowment for the Arts in 1981, and must be seen in an evolutionary context. That report defined the field, indicated its importance in terms of social and economic benefits, and demonstrated the public utility of environmental design research. No doubt problem-solving priorities will shift over the years as different ways to structure the field emerge in the scientific literature and the environmental professions. The current book represents a broad consensus of where things stood at the end of 1981 and where they should go in the future. It assumes the utility of environmental design research, and goes on to address cutting-edge issues. Continued progress in this field may be measured against this reference point.

PROCEDURE

From late 1979 through mid-1981, the Environmental Design Research Association conducted an organizational effort to develop a representative and comprehensive statement about research needs for the 1980s. Funding for the project was made available by a Design Exploration/Research grant from the National Endowment for the Arts (NEA), with matching contributions from EDRA and from the University of Wisconsin–Milwaukee.

The project was conducted by the authors under the direction of the Board of Directors of EDRA. Throughout the process, careful consideration was given to ensuring representative input from the entire field of environmental design research. The EDRA membership and other leading scholars were invited to offer input in several ways: a letter survey was conducted of leading researchers, scholars, research administrators, and practitioners in the field; workshops were held at annual EDRA conferences; position papers and mini-research agenda were received and incorporated; an Interorganizational Task Force on Human-Environment Research and Applications was invited to offer input from its multidisciplinary perspective; and various drafts of this monograph have been reviewed by the Board of Directors, a special Task Force on Research Agenda, and a Research Agenda Review Committee of EDRA. Details of the procedure are presented in the Appendixes.

AUDIENCE

The book is intended primarily for two audiences: for researchers, academics, and professionals working in the social sciences and the environmental professions dealing with issues of environment-behavior-design relations; and for the organizers, supporters, and teachers of research in professional associations concerned with environment-behavior issues, public and private research granting agencies and foundations, university and private research organizations, undergraduate and graduate faculties and training institutions, and all levels of government where policy makers use the results of environmental design research. The book is thus intended for researchers, professionals, teachers, and administrators in the following fields: architecture, landscape architecture, urban design, urban and regional planning, interior design, human factors, environmental psychology, environmental and urban sociology, social and behavior geography, housing and urban affairs, urban anthropology, environmental gerontology, child development and early childhood education, and environmental education. Even allowing for duplication of membership lists of the environment and behavior sections of the associations in each of these fields, the subscription lists of the major journals in the field, and the EDRA membership list, a conservative estimate is that close to 5,000 people are involved in this field—in academe, in the professions, in government, and in business. We hope it will also have value for others outside the field who are concerned with the quality of the human and physical environments.

ORGANIZATION OF THE BOOK

The book is organized in three parts. Part I provides a synthesis of the field. Chapter 1 offers a way of defining the field. Chapter 2 sketches some of its history. Chapter 3 indicates the most important underlying assumptions. And Chapter 4 offers a conceptual framework for organizing and identifying major research areas.

Part II outlines areas, topics, and issues needing immediate and sustained research attention. Chapter 5 deals with the need for improving theories of environment-behavior-design relations and exploring other critical theoretical issues. Chapter 6 looks at substantive issues related to different types of environmental settings from the macro to the micro scale. Chapter 7 explores issues related to the great diversity of different groups of people. Chapter 8 treats issues related to socio-behavioral phenomena ranging from physiological responses to the environment to cultural values toward the environment. Chapter 9 discusses pressing process issues of research, application, and research utilization, and of environmental problem solving, decision making, planning, and design. Chapter 10 examines a number of important issues about the contexts for environmental design research, both in its pure and applied aspects.

In these chapters, no attempt is made to review the current research literature. Examples of research are cited and references given for the avid reader. In the presentation of research directions, a rigid hierarchy of priorities was also deemed inappropriate for the field. The monograph is a statement of important directions that the field should be taking in the 1980s and 1990s, with different research centers, firms, and agencies focusing on areas and topics fitting their contextual needs, constraints, and intellectual-professional interests.

Part III suggests some of the strategies necessary to see the research directions suggested in the book implemented through legislation, granting agencies and foundations, research institutes, and education (Chapter 11).

BIASES

While the content of the monograph has been influenced by the ideas and suggestions of many people and associations, it is a position statement representing the opinions of those who sent in letters and wrote position papers, the EDRA Task Force, other contributors, and the authors. It is not to be construed as the policy of its sponsors, the National Endowment for the Arts, nor of every member of EDRA. The fact that we have devoted our careers to scientific research, both basic and applied, to applications to environmental problem solving, and to university teaching means that there is nothing neutral about our attitudes toward science and applications to environmental problem solving. As Herbert Simon stated in his National Academy of Science's report to the National Science Foundation in 1976, "We believe that scientific knowledge provides an essential foundation for modern society and that the preservation and further progress of that society calls for the vigorous pursuit of science, both basic and applied. The social sciences [and the environmental design research sciences] are an integral part of total scientific endeavor, and they must and can be pursued with objectivity, respect for evidence, and intellectual sharpness."

ACKNOWLEDGMENTS

This work has been developed under a design research grant to the Environmental Design Research Association from the National Endowment for the Arts, Design Arts Program (Gary T. Moore and Sandra C. Howell, Project Directors), with matching contributions from EDRA and from the senior author's host institution. We would like to thank Michael Pittas, former Director of the Design Arts Program, and Charles Zucker, then Associate Director, who supported the development of this work. Thanks are due also to Anthony James Catanese, past Dean of the School of Architecture and Urban Planning at the University of Wisconsin–Milwaukee, for making the research facilities, services, and supplies of the School available for this project, and to Francis Ventre of the National Bureau of Standards and Frederick Krimgold of the National Science Foundation for their initial encouragement and assistance throughout.

Members of the Task Force on Research Agenda, Research Agenda Review Committee, and Boards of Directors of EDRA during the 1979-82 period have contributed to this work by developing position papers and by comment and review. In particular, we would like to acknowledge the conceptual contributions of Irwin Altman, John Archea, Michael Brill, Robert Gutman, William Ittelson, John Zeisel, and Ervin Zube. A special thanks to Daniel

Stokols, the Chair of the Research Agenda Review Committee, who contributed so much by idea, comment, and example throughout the last year of this project, to Robert Gutman, the 1981-82 Chair of the EDRA Board of Directors, who contributed a large portion of Chapter 10, and to Robin Moore, the 1983-84 Chair of the EDRA Board of Directors, who revised the Synopsis. The monograph also owes much to the ideas and counsel of Ernest Alexander, Robert Bechtel, Richard Bender, John Boal, Sidney Brower, Richard Chenowith, Uriel Cohen, John Eberhard, John Habraken, Gary Hack, J.B. Jackson, Jonathan King, Paul Laseau, Robert Marans, Stephen Margulis, William Michelson, William Mitchell, Arthur Patterson, John Rahaim, George Rand, Amos Rapoport, Janet Reizenstein Carpman, Leanne Rivlin, Robert Shibley, William Sims, Paul Taylor, and Joachim Wohlwill. We would also like to thank the following people who provided other agendas: Frederick Krimgold of NSF, Michael Pittas of NEA, Roger Schluntz of the Association of Collegiate Schools of Architecture (ACSA), James Snyder of the Architectural Research Centers Consortium (ARCC), John Dixon of *Progressive Architecture*, and Willo White, the Executive Officer of EDRA.

A very special thanks to Annabelle Sherba at the University of Wisconsin–Milwaukee who typed the many drafts of this book and coordinated all aspects of production. We feel very fortunate to have her associated with this project. Special thanks also to Ellen Bruce and Patricia Gill of the Environment-Behavior Research Institute at UWM who each outlined and drafted a chapter, to Donna Duerk of MIT who provided research assistance, to Susan Fulton, Karen Kruming, Naomi Leiseroff, Elizabeth Kubale Palmer, Marleen Sobczak, and Jeanne Zagrodnik of UWM who assisted in analyzing the letter surveys and in compiling the references, to Anthea Van Skoyk of the UWM Department of English who gave the manuscript a thorough copy editing, to Mary Eichstaedt of the UWM School of Architecture and Urban Planning who helped greatly in the final typing, to Aniruddha Gupte who put the final touches on the illustrations, and to Kathryn Simmons who helped compile the index. My appreciation and sincere thanks to all.

Finally, we would like to thank the many people inside and outside the Association who have contributed their thoughts and ideas, all of which have been reviewed carefully, and all of which have greatly influenced this document.

Gary T. Moore
Milwaukee, Wisconsin

Synopsis

Environmental design research is the study of relations between people and their surroundings. The purpose of the field is to produce information that can be used to improve the quality of life through environmental policy, planning, design, and education. This policy document from the Environmental Design Research Association (EDRA), summarizes the findings of the EDRA Future Directions Study. Recent developments in the field of environmental design research are examined, a framework for organizing the field is suggested, unresolved and new issues are highlighted, and research directions and priorities for the decades ahead are proposed.

Critical world problems that bear on the quality of the environment are not unique to the 1980s. People have always been faced with wars, famines, natural disasters, and disease. Each generation is faced with its own unique brand of problems, conditions, and circumstances. A number of social, political, and environmental conditions face the world today and provide the context for research on environment-behavior issues and applications to improving the quality of the environment. These issues include limitations in global resources, the continued deterioration of the environment, dramatic social changes brought about by rapid technological development, and the volatile nature of current political and economic systems for the allocation of resources.

Limitations in food, water, energy, land, and the materials of production, along with the ever increasing pollution and deterioration of the environment, define the limitations in which we must try to provide an improved quality of life. These conditions affect the way people respond to the environment when making decisions. The dramatic social and cultural changes affecting every aspect of our lives define in part how people relate to their environment, and are major modifying factors in that relationship. With the dramatic changes now affecting cultures, sub-cultures, and life styles, the existing physical environment is no longer able to satisfy its users in the same way as before. Women are no longer willing to stay at

---

This synopsis owes much to the editorial work of Robin C. Moore and members of the EDRA Board of Directors.

home and take care of the house and children, but are demanding their rightful place in society as equal partners with men in the work force and in meaningful decision making. Older Americans are demanding social and economic security and an environment that will continue to provide challenges for successful aging.

Major population shifts are happening in the country, away from the industrial northeast and midwest, away from central cities, and to the sunbelt and the suburbs. While there are few surprises about this continuing phenomena, it places a grave responsibility on society to continue to provide life supporting environments in those areas where the economic base may be withering, while providing for the needs of areas impacted by new populations.

The coming of the postindustrial society is having massive effects on the work place and on social change. Present industrial work places will no longer be operating under the same assumptions and conditions that they once were. Worker productivity oscillates, and has been at all time lows in the developed world. Meanwhile, there is evidence that secondary industry (manufacturing) and its direct services (light manufacturing and assembly) may be experiencing a rapid and profound decline, and will go the way of primary industry (farming, fishing, and mining). Instead of ninety-five percent of the workforce being engaged in primary industries, in the United States today less than five percent manage to do these things for us. A similar and profound shift from secondary to tertiary industries: (automation, information processing, and services) is underway. There are few who doubt this trend will continue, and will have profound effects on what we call the quality of life, and on demands people will place on the physical environment for new opportunities like leisure activities, recreation, and re-industrialization in the new modes. It is not only important to recognize the impact of these changes, but also to understand the process of cultural change itself as a factor mediating the environment-behavior relationship.

The political and economic context of today's societies must also be understood if research and professional interventions are to be effective in helping to solve environmental problems. It is the economic and political context that provides environmental researchers, design professionals, and environmental policy makers the conditions and constraints for improving the quality of the environment and, thereby, the quality of life.

ENVIRONMENTAL DESIGN RESEARCH: ENVIRONMENT— BEHAVIOR RESEARCH AND APPLICATIONS

In the zeitgeist of the sixties the need for a more socially responsible approach to the planning and design of the environment was identified. To meet this challenge, a new multidisciplinary field emerged from the more traditional disciplines. This field has been labeled environment-behavior studies, environmental psychology, and environmental sociology—here called *environmental design research*.

Research on human behavior and the environment can be traced back to the turn of the century. Sustained programs of research and publication began in the fifties. Passage of the community Mental Health Center Act (1963) provided an early stimulus to the emerging field. A massive building program was underway to meet the needs of a highly vulnerable group (mental patients). The collaboration of design and mental health professionals on the programming of new facilities meant that both groups became more sensitized to the effects of the physical environment on human behavior. A pair of national conferences in the mid-1960s sponsored by the U.S. National Institute of Mental Health gave initial visibility and structure to the field. Publication of the first directory of "Behavior and Environmental Design" followed in 1965, and the first collection of work was published in the *Journal of Social Issues* in 1966.

The Environmental Design Research Association (EDRA) was formed in 1968. EDRA is the oldest and largest body in the world devoted to environment-behavior-design concerns. The organization sponsors annual conferences, publishes research proceedings and monographs, cooperates in the publication of *Environment and Behavior* and the *Journal of Architecture and Planning Research*, collaborates with publishing houses in the preparation of two series of books and works with related social science and environmental associations to support efforts toward environmental quality at all levels of decision-making.

VALUES AND GOALS

The scope of environmental design research draws from concepts and methods in both science and art, and the professions. It assumes that science is not value-free. The field is oriented to all levels of human experience (from physiological responses to social and cultural phenomena) at all scales of the everyday physical environment (from the micro scale of interiors to the macro scale of regions). In order to respect the integrity of the transactions of people and events in everyday settings, environmental design research adopts a variety of scientific methods from the social and behavioral sciences and relies heavily on the rigorous application of exploratory, descriptive, quasi-experimental, and field research methods. The bottom line is a respect for environmental justice and a call for the redress of injustices in the form of inaccessibility, exclusion, or unequal distribution of environmental resources and amenities. The field learns from and contributes to public policy formation, urban and regional planning, civil and environmental engineering, urban design, landscape architecture, architecture, interior design, and industrial, graphic, and product design.

The field has had an impact on a wide range of issues. They include: national housing policies; planning and design policies to reduce crime and vandalism; design standards for human safety; revised national standards for handicapped accessibility; cost-benefit economics of worker satisfaction and office productivity; and planning and design guidelines for military housing, community recreation buildings, and children's environments.

THE "FUTURE DIRECTIONS" STUDY

From 1979 through 1981, under a grant from the National Endowment for the Arts Design Research Program, the Environmental Design Research Association conducted an effort to define and communicate environment-behavior research directions for the 1980s. A survey of research needs and proposed directions was conducted among members of the Association, other scientists, and practitioners involved in environment-behavior studies and applications. Position papers were prepared by an EDRA Task Force, suggestions were made by an Interorganizational Task Force on Human Environment Research and Applications, other agendas in the field were reviewed, and oral comments were received from EDRA members.

The study provides a synthesis of the field to guide future research and environmental applications. It presents a set of research directions for policy makers in public and private sectors: federal research agencies, research sections of mission agencies, private foundations, industrial and corporate sponsors of research, scientific and professional associations, university departments, research institutes, research training establishments, and private research and research utilization organizations.

THE SCOPE OF THE FIELD

As a field environmental design research can be conceptualized along six major dimensions: (1) *places* used by people within the overall environment; (2) impacted and vulnerable *user groups* in those settings; (3) physiological, psychological, behavioral, social, and cultural *phenomena and responses* to the environment; (4) evolution of environment-behavior relationships over *time*; (5)

development and testing of *theories and conceptualizations*, and (6) *processes of utilizing* research-generated knowledge in environmental policy, planning, design, and education.

The task of environmental design research, therefore, is to investigate, document, and explain the relations among these six dimensions at all levels of human experience and all scales of the physical environment, and to indicate implications for improving the quality of life through better informed policy, planning, design, and education.

The remainder of this synopsis presents the most important findings and recommendations on priority issues needing immediate and sustained attention. Recommendations are also made for improving major programs of research on these issues. The full rationale is given in the text that follows.

BASIC RESEARCH: TOWARD THE CONTINUED DEVELOPMENT AND TESTING OF EXPLANATORY THEORIES

After twenty-five years of sustained research the field is "data rich." Theoretical orientations, frameworks, and models have advanced the general conceptual basis of the field and its relevance to critical problems of the human environment. Priority still needs to be given to basic research that further develops, articulates, and tests theoretical orientations, frameworks, models, and especially middle range explanatory theories. As Nobel Laureate Herbert Simon said, "Society supports basic research because of a belief that fundamental advances in knowledge will lead to important practical applications, and conversely, that advances in practical knowledge and technology rest on the foundation of basic knowledge. The history of the past two or three hundred years provides a mass of evidence to support this belief" (National Academy of Sciences, 1976, p. 20).

Conceptual refinement of the notion of environmental quality and its relationship to the quality of life is needed. This includes the specification of an empirical taxonomy of settings and dimensions that matter most to human health and welfare, and the clarification of critical outcome variables such as satisfaction, health, productivity, and development. Fundamental investigations are also needed concerning the limits of the impact of the environment on behavior, and the methods people use to make trade-offs between subjective and economic satisfactions. Many of these issues need clarification from cross-cultural, cross-national, and life-span points of view.

The resolution of these conceptual issues will contribute to the development of a more integrative and theoretically focused characterization of the interactions between people and the physical environment. Such a coherent characterization will also have considerable benefits for the professional application of environmental design research.

PLACE RESEARCH: SETTINGS FOR HUMAN EVENTS

Research issues organized by places, settings, and different types of environments are emerging as a new area of substantive environmental design research. Attention needs to be focused on the factors of physical settings, human experience, and human cognition that contribute to the essence, history, and management of places.

Major problems facing our environment occur at urban and regional scales. While the field has concentrated on building-scale issues, environmental design research should also focus, therefore, on a variety of large-scale natural settings and issues such as landscape assessment, aesthetic values, the history of the evolution of the designed landscape, and on theories that account for existing data on perception and behavior in different large-scale environments.

Intermediate-scale places that need additional attention in the future are workplaces; housing for low-income, elderly, handicapped, and ethnic populations; and understudied everyday places such as shopping centers, parks, and community facilities. Places that have been affected by rapid technological changes like hospitals, research laboratories and scientific centers also need study.

Other topics include the smallest scale of interior design, building materials, and objects—areas of study which have been consistently neglected in the field—impacts of the environment on health, safety and security, the toxic effects of building materials and accident prevention. More knowledge is also urgently needed about human reaction to sensory stimuli in the luminous, sonic, and thermal environment.

USER GROUP RESEARCH: PEOPLE IN PLACES

Past research has focused on the needs of the elderly, children, and the handicapped. The interrelationships of these groups must be recognized within larger economic, political, and social frameworks. Such an approach acknowledges design topics relevant to each group while integrating their various environmental requirements into a new order of five larger groups on which research is needed: (1) groups persuing new lifestyles; (2) groups with special life-cycle needs; (3) politically under-representated groups; (4) other groups that are vulnerable to the environment, and (5) "ordinary people."

An area of user group research should focus on changing life cycle and lifestyle requirements. Much of the environment is designed for only one segment of the population in one stage of the life cycle: middle income, middle age, white Anglo-Saxon. Continued research should be devoted to aging, as this represents the fastest growing segment of the population. Other research should focus on the environments of children and adolescents, and of those people in intermediate stages of the life cycle.

A second group needing research attention are those with least control over their day-to-day environment, including institutionalized persons, those with physical and mental disabilities, and those incarcerated in prisons or mental institutions, and, to a degree, all politically under-represented users, cultural minorities, and vulnerable groups.

Finally, at the other extreme, much of the Western world is comprised of ordinary everyday environments occupied by "ordinary" middle-class citizens who have been overlooked in design research simply because they do not stand out as a particular, identifiable group.

SOCIO-BEHAVIORAL PHENOMENA: PHYSIOLOGICAL TO CULTURAL RESPONSES TO THE ENVIRONMENT

Since its inception, a major focus of empirical research in the field has been behavioral and socio-cultural responses to environments. While the field of environment-behavior studies is experiencing a shift towards issues organized by place, there are still a number of social and behavioral research issues needing attention. The research needs in this area can be conceptualized in expanding circles from inner, personal responses to external, socio-cultural responses to environments.

Research on internal physiological responses to the environment could usefully focus on major potential health hazards of the environment, including from new building and information technologies. While much is known about psychological responses to the environment (e.g., environmental cognition, crowding, stress), the needs of the design-oriented environmental professions still call for expanded research, especially on environmental perception, user imagery associated with different designed environments, meaning and symbolism, and emotional responses to different environments.

External behavioral responses to the environment deals with those behaviors of the individual, social group, or larger society and culture that are manifest in cultural artifacts. These behaviors and their manifestations (e.g., house form as a manifestation of culture) are typically more visible than physiological and psychological responses, and have therefore been more easily and more often measured. Research now needs to be focused on behavioral issues of environmental choice, the locus of control in the environment, effects of density and crowding in different age and life-cycle groups, the relation between job satisfaction and performance, and environmental factors related to all of these responses.

At the social level, priority research issues include examination of social class in relation to special factors in the environment, studies of large samples of populations over time, social group differences in response to the environment, and change and adaptation over time.

Finally, priority issues at the cultural level include the relation of different social norms, family and social organizations, and institutional structures on cultural values with respect to the environment. Another major topic concerns cross-cultural research and cross-cultural perspectives on environment-behavior concepts developed in contemporary Western cultures. Equally important, the ethnic and cultural composition of the United States is changing rapidly, as seen in the projected increase in Hispanic populations, the increasing pluralization of the country, the increase in black occupancy of urban areas, and massive population shifts around the country, all of which portend the need for environmental design research to study environments in relation to culture.

PROCESS RESEARCH: TOWARD GREATER UTILIZATION OF RESEARCH IN THE PROFESSIONS

Environmental design research is problem-centered and action-oriented. A central set of issues relates to the use of knowledge about environment-behavior relations in the professions dealing with environmental policy, planning, design, and education. In environmental policy planning important research issues about knowledge utilization include studies of how environmental design research information is used in policy decisions in public and private sectors, tests on the applicability of models of public policymaking for the application of environment-behavior knowledge in other fields, and continued investigations of the factors contributing to the successful use of research.

Concerning the responsiveness of environmental design to environment-behavior research information, priority issues include: investigations of the strengths and weaknesses of different design and decision-making methods; assessment of different facility programming methods, including the validity of different information gathering and user participation techniques; identification of the critical aspects of user needs that effect design solutions; methods of presenting design criteria and design guidance in ways that do not constrain design solutions needlessly; development and testing methods incorporating cost-effectiveness; methods of improving participatory design methods and making them more feasible in conventional practice; and the responsiveness of computer-aided design to environment-behavior and direct user impact.

With regard to environmental assessment and evaluation, pressing research issues include strategies for including environmental evaluations in the normal design-build cycle for both large public clients and smaller private clients.

Priority research on environment-behavior research itself includes: comparative assessment of different research procedures and designs for capturing the essence of environment-behavior events; further development and critical analysis of the advantages and appropriate applications of both quantitative and qualitative approaches—including experimental, quasi-experimental, archival, and action-research; and the development of reliable research tools and techniques that can be applied in professional decision-making contexts.

Research priorities in the area of knowledge transfer between disciplines include: studies of data management and translation into policies, planning regulations, and design standards; strategies for dissemination of research findings to public and private decision-making communities; and studies of media and their communication effectiveness. Aesthetic criticism, for instance, needs to view the built environment as a human habitat as well as aesthetic objects in the landscape. The social commitments of design and planning need reinforcement. Conversely, the rhetoric of social responsibility must embrace aesthetic theory. Design based solely on formal rules of composition or solely on political ideology puts unnecessary distance between designers and the public they serve.

RESPONSES TO SOCIAL AND TECHNOLOGICAL CHANGE

The last twenty-five years have brought about dramatic changes in lifestyles in many parts of the world. In Western societies, the struggle of women for new roles and responsibilities has been a major development with resulting changes in family structure, the role of the family in society, and changes in marriage mores, in styles of parenting, and in female and male roles. An important task of environment-behavior research is to help adapt the environment to these social changes.

Limitations in food, water, energy, land, and materials for production, along with increasing pollution and deterioration of the environment, define the arena in which an improved quality of life must be provided. The dramatic changes brought about by rapid technological development, now affecting all cultures and lifestyles, means that the physical environment will no longer be able to satisfy its users in the same way as before. Women are no longer willing to stay at home as caretakers of house and children. Older Americans are demanding social and economic security and environments that will continue to provide challenges in later life.

Major population shifts are happening away from the industrial northeast and midwest, away from central cities, towards the sunbelt and suburbs, placing added responsibility on society to continue to provide life supporting environments in those areas where the economic base is declining. Secondary industry (manufacturing and assembly) is experiencing a rapid and profound metamorphosis. A similar shift from secondary to tertiary industry (automation, information processing, and services) is taking place. There are few who doubt these trends will have great effects on the quality of life and our environment.

PRIORITIES FOR IMPROVING THE IMPACT OF ENVIRONMENTAL DESIGN RESEARCH

A major concern for any relatively new field, particularly one at the interface of several disciplines, is to find ways to produce higher quality research. To this end, six objectives are advanced:

1. To identify major *problems* requiring environmental design research . . . conduct broad surveys of research needs in the environmental professions, and encourage the periodic and systematic review of portions of the literature on research and research applications.
2. To increase the *awareness* of the needs and utility of environmental

design research...evaluate and communicate examples of effective environmental design research, initiate media campaigns to inform the general public and the professions about available new knowledge, develop on-going programs for the dissemination of research findings, and develop programs to increase recognition, incentives, and public awareness of successful environmental design research.

3. To influence *legislation* impacting the conduct of environmental design research...monitor legislative programs, draft model legislation, and actively support bills of interest to the field.

4. To influence patterns of *funding* for environmental design research...identify the major potential funding sources for environmental design research, better understand the political process of decision-making surrounding research, work actively to increase funding, and identify, encourage, and communicate the results from programs of research that have made a difference to the field and to applications, including large programs of research and research that has a high probability of contributing to the solution of major human environmental problems.

5. To improve *education* in environmental design research...encourage more, expanded, and better programs of graduate research training, and new and expanded programs of continuing education initiated by the professions, and encourage the use of research and the translation of research into environmental problem-solving at all levels of education.

6. To further *communication* within scientific and professional associations...launch an *interorganizational* effort to develop a centralized lobbying and political action coalition.

Environmental design research is committed to a more socially responsible and humanistic approach to environmental design and to research bearing on the quality of the environment.[1] Unheeding the call in the 1960s for more socially responsible planning and design, twenty years later, despite pockets of change, much of our physical environment has become more alien, less responsive, less healthy, less supportive of community and social interaction—in a word, less humane. The new formalism shaking the design professions, and the bureaucratic and economics-for-the-developer concerns that are eating away at the consciousness of the planning professions, are a retreat from ethical and moral responsibility, and sacrifice the concerns and dreams of common people for an elitely defined aesthetic and economic movement of the few.

Environmental design research is committed to the fight for social justice, to diversity, to the emergence of community and public life, to direct participation and empowering people to control and be responsible for the design, construction, use, and maintenance of their environment, to improved research and the use of accumulated knowledge about people's needs, desires, perceptions, and behaviors, and to truth in presentation.

This book, *Environmental Design Research Directions: Process and Prospects* is a modest attempt to contribute to these goals by examining recent developments in the field, suggesting a framework for organizing the field, highlighting unresolved and new issues, and presenting a set of research directions and priorities for the future.

NOTE

1. Our field and the nation has recently lost one of our most respected and revered colleagues, Donald Appleyard, through an environmental accident that even his best research and environmental design efforts could not prevent. We mourn his loss. The final section is based heavily on one of the last pieces he and his colleagues wrote, "A Humanistic Design Manifesto," June 1982. (Since this writing, we have lost two more revered colleagues—Kevin Lynch and Thomas Byerts. Their passing is a sad commentary on the maturity of our field. We miss them all greatly.)

ENVIRONMENTAL DESIGN
RESEARCH DIRECTIONS

PART I
THE NATURE OF
ENVIRONMENTAL DESIGN
RESEARCH

1

WHAT IS ENVIRONMENTAL DESIGN RESEARCH?

The definition of environmental design research as basic and applied research and research utilization dealing with environment-behavior relations and applications to improving the quality of life through better informed environmental policy, planning, design, and education.

Science is one of the powerful tools that humans have developed for deepening their understanding of the physical, biological, and social world around them; for increasing their ability to act intelligently and effectively in that world; and for enabling them to estimate the environmental and social consequences of their actions.

National Academy of Sciences
*Social and Behavioral Sciences in the
National Science Foundation, 1976*

While the natural and engineering sciences have long been closely integrated into architecture, the material sciences into interior and product design, horticulture into landscape architecture, and the policy sciences into urban planning, it is only since the 1950s that the social and behavioral sciences have been accepted into the environmental professions. This limited acceptance is surprising in light of the frequent allusions to human scale and human use in the environmental design and planning literature. There is a widespread myth among designers that behavioral science information will somehow constrain their art, and among planners that the only reality is policy and economics. There are some environmental professionals who ignore human use while there are those who would like to know more, and would use research information more

3

if the structure of the environmental professions were to reward it. Conversely, the social and behavioral sciences operate on the belief that informed planning and design results in enhanced creativity, but have had great difficulty communicating research findings to the community of environmental professionals.

Since the 1960s, a growing number of researchers from various disciplines have directed their attention to connecting the environmental professions with a wide range of social and behavioral sciences. These efforts have had significant impact on the shape and effectiveness of the built environment. A new literature is on the shelves of designers' offices, and references to the intended human effects of design solutions are becoming more common in the presentations of designers to their private clients and planners to public agencies and communities.

DEFINITION

Environmental design research (EDR) is the study of the mutual relations between human beings and the physical environment at all scales, and applications of the knowledge thus gained to improving the quality of life through better informed environmental policy, planning, design, and education. Environmental design research focuses on the interdependence of physical environmental systems and human systems, and includes both environmental and human factors. The field also includes studies of the political, social, and economic context of research, studies of environmental intervention, the processes of research, decision making, planning, and design, and studies of communication, research translation, and information dissemination.

A host of designations for this field has been employed: environment and behavior and environment-behavior studies, but also environmental psychology, architectural psychology, ecological psychology, social ecology, environmental sociology, environmental perception, and so on. Environmental design research, as used here, encompasses large portions of each of these other terms. Many of these designations have been developed within the context of one or another parent fields like environmental psychology within psychology, environmental sociology within sociology, and so on. This has led to the development of several professional associations (for example, EDRA, the environmental sections of the American Psychological Association, American Sociological Association, American Institute of Architects, and others). Some of these terms have evolved

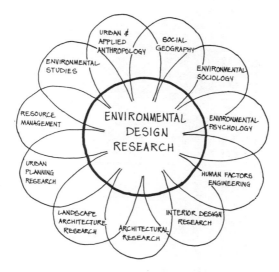

Figure 1 Environmental design research (EDR) is a confluence of many parts of the social sciences and environmental professions.

around a particular approach or theorist, like "ecological psychology" around the work of Barker and his students. The interdisciplinary nature of the field (see Figure 1), the dialectic unity between environment and behavior, and the interest in improving the quality of life through the physical environment have meant that two designations—"environment-behavior studies" and "environmental design research" seem to have supplanted all others.[1]

Many of these designations have implied pure research endeavors without specific attention to the alleviation of environmental problems or the application of knowledge gained to the environmental professions. As illustrated in Figure 2, environmental design research is concerned with the gamut of research and applications—from basic research and the construction of explanatory theories of environment and behavior to applied research and applications to the environmental professions. Many of the questions addressed in the field are influenced jointly by considerations of theory and of practical utilization to environmental problems. It must be made clear that what is here termed environmental design research also encompasses studies of any applications to environmental policy, planning, and education. Thus *design* is used in the broad sense of the invention and disposition of the parts of the environment according to a purpose and plan.

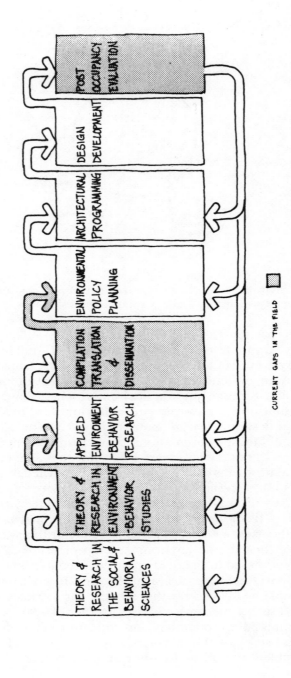

Figure 2 The gamut of research and applications for people-in-environments (based on Moore, Research, design, and evaluation, 1979[c]).

6

Villecco and Brill (1981) offer these illustrations of what may and may not be considered environmental design research.[2]

1. Environmental design research can be distinguished from other kinds of research by its emphasis on the relationship between people, the physical environment, and implications for the quality of life. Research on the seismic resistance of alternative structural systems for buildings would be considered building science or structural engineering research. Research on people's cognitive images of the stability and safety of various kinds of structures, and their associated willingness to inhabit different buildings in seismic zones, is environmental design research. The first of these examples might be questioned since every dimension of the design process has implications for human safety and well being. Environmental design research has chosen to highlight the behavioral and social dimensions of planning and design and the physical contexts for human experience and behavior, since these issues have been underemphasized by both the environmental professions and the social sciences.

2. Environmental design research can be distinguished from design practice. If an architectural firm were to propose that they research new lighting systems for a particular project, this would not be environmental design research, even though the effort might include a literature search and site visits to similar projects to expand the firm's knowledge base. The purpose of such a venture is to apply existing knowledge to a single building, rather than generate new knowledge. Research on the impact of artificial versus natural light on job satisfaction and productivity is environmental design research directed toward the advancement of the field.

3. Environmental design research deals with the art as well as the science of design and environmental problem-solving. Environmental design research strives to generate knowledge useful to environmental policy, planning, and design. The role of art and of formalist controversies within the design professions today is directed also at improving the quality of the designed environment. If inquiry into the artistic elements of design is structured into a rational and explicit research project, then it is environmental design research.

Environmental design research includes basic research, applied research, and research applications. First, *basic research* is the generation of knowledge and the discovery of phenomena, processes, and systems that are potentially significant to understanding environment-behavior interactions and to developing a general theory of people, environment, and environmental design. Basic research, unlike applied research, does not need to be justified on the basis of specific problem solving. Its applications may be unclear or unpredictable but it is critical to the overall development of the field, to environmental design theory, and to a more responsive environment. Second, *applied research* is concerned with answering specific questions relative to an immediate social policy context and problem. In applied research the question and the form of its investigation are defined by the needed application, the client's goals, and the time frame. Third, *research applications* are the compilation and translation of findings from both basic and applied research into specific environmental policies, plans, or designs. The recent interest in the development of design guides, environmental programs, and information dissemination are examples of research applications. The application of research to practice is critical; it is not environmental design research, however, unless it is proposed to study those techniques which are most effective in communicating research results to policy or professional applications.

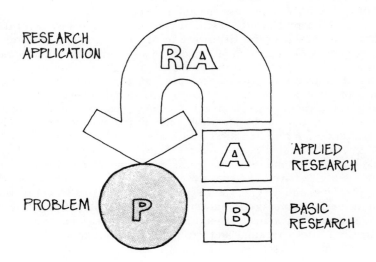

Figure 3 Environmental design research is multilevel.

NOTES

1. A recent review of environmental psychology (Russell & Ward, 1982) has defined environmental psychology as a branch of psychology "concerned

with providing a systematic account of the relationship between person and environment." Their definition sees environmental psychology analogous to what they consider to be developmental psychology, that is, providing a unique perspective on strictly psychological processes. It may be disputed that this is the intention of developmental psychology; see the variety of multi-disciplinary contributions in the major journal, *Child Development and Behavior*. They further differentiate "environmental psychology" from "ecological psychology," "human ecology," "eco-cultural psychology," and "behavioral geography." Finally, they stop short of equating environmental psychology with "the larger field of environment and behavior." Rather, they have defined environmental psychology as "the area of *overlap* between the two *larger* disciplines, psychology and environment and behavior. This may be the first time that other disciplines have defined themselves in terms of the larger field of environment and behavior research, or what we are here calling "environmental design research"—a sure sign of the final emergence of the field. In writing this book and reviewing reviews from other members of the field, the question constantly arose about what name to adopt. While there are good reasons for referring to the field as environment-behavior studies (to witness its truly multidisciplinary character and not ally the name primarily to any one parent discipline), we will refer to it as environmental design research in this volume to acknowledge the design content of much of the discussion and the fact that the book is produced under the auspices of the Environmental Design Research Association. Similarly, the forthcoming *Handbook of Environmental Psychology* is titled to reflect its psychological flavor and its connection to the American Psychological Association. The three terms—environmental design research, environmental psychology, and environment-behavior studies—are used interchangeably in most of the field, and in this book.

    2. These four examples are paraphrased from Villecco and Brill (1981).

2

HISTORY AND IMPACT OF THE FIELD

The origins of the field from both the social science and the environmental professions. The importance of the early leaders (Lewin, Wright, Firey, Barker, Jackson, Osmond, Sommer, Ittelson, Proshansky, Hall, Rapoport, Lynch, Gans, Altman, Stea, Kates, Wohlwill, and others). The beginning of the major organization (EDRA) and the major publications (*Environment and Behavior* and other journals). The utility of the field, with examples from office environments, slums and social insecurity, public housing policies, vandalism in public schools, safety, access to buildings for the visually and physically handicapped, and others.

A HISTORICAL SKETCH[1]

Research on the human perspective in environmental design, planning, and architectural policy, and research on the physical environmental context of behavior may be traced back at least to the turn of the century. Systematic research on environment-behavior issues and applications began in the early 1950s. Prior to that time, four ground-breaking and very influential works were: Firey's (1945) studies of city symbolism in sociology; Lewin's (1946) studies of child behavior as a function of the total situation; Wright's (1947) geographical study of changing mental conceptions of the environment; and Festinger, Shacter, and Back's (1950) study in social psychology of the development of informal social groups in university housing as a function of design factors.

A series of well-known studies was initiated during the early 1950s that has continued to have a marked impact on the field.

Among these were studies of the ecological psychologist Barker on the relation of behavior to characteristics of small towns in the United States and England (Barker & Wright, 1955).[2] Jackson's writings in *Landscape* through the 1950s had a profound influence on many scholars in the field (Jackson, 1951-1968; Zube, 1970). In Canada, Osmond (1957), a research-oriented psychiatrist, published a paper on psychiatric ward design, and Sommer, a social psychologist, completed a study of geriatric wards demonstrating for the first time the concept of territoriality (Sommer & Ross, 1958). At about the same time, Ittelson, Proshansky, and Rivlin (1970), began a research program on the influence of ward design on patient behavior in mental hospitals, and the anthropologist Hall (1959) published his now-classic books on the silent language and hidden dimensions of behavioral and perceived space in different cultures.

During the 1960s and 1970s, the field emerged as one of the fastest growing areas of both psychological and architectural research, and witnessed major contributions from a variety of other disciplines from human factors to urban anthropology and from interior design to public administration. Classic works published during that time include the architect and anthropologist Rapoport's (1969) book on the relation of culture to house form,[3] Lynch's (1960) book on the image of the city, the series of bibliographic volumes edited by Larson (1965), sociologist Gans' (1959) study of the life space, housing, and neighborhoods of immigrant groups, Altman's studies of the ecology of isolated groups (Altman & Haythorne, 1967),[4] and the environmental psychologist Stea's (1969) studies of cognitive mapping.

Collective interest among psychologists, geographers, sociologists, and architects became evident with the publication of the first directory of "Behavior and Environmental Design" (Studer & Stea, 1965) and first journal issue devoted to this new area, the *Journal of Social Issues*, "Man's Response to the Physical Environment" (Kates & Wohlwill, 1966). Fueling interest in the field was the growing public concern for the quality of the environment, as outlined in the U.S. Environmental Policy Act of 1969. The result of these forces has been widespread research activity in universities, industry, government, and the professions, and the beginning of the application of findings in government policy making and professional practice.

As an expression of these developments, a group met at the Massachusetts Institute of Technology in 1968, immediately after the first International Conference of the Design Methods Group. Representatives from both the Design Methods Group and the Architectural

Psychology Newsletter met to create a new professional association, the Environmental Design Research Association (EDRA).
EDRA held its first annual conference in 1969 and published its first
annual *Proceedings* the same year. EDRA is the oldest and largest
organization in the world devoted to environment-behavior-design
studies and applications to improving the quality of the physical
designed environment. The major research journal in the field, *Environment and Behavior*, began in 1969 and has been published in
cooperation with EDRA since 1980.

In 1974 the American Psychological Association formed a Task
Force on Environment and Behavior and began a joint Division of
Population and Environmental Psychology in 1976. Other scholarly
and professional associations have since created sections focused
on issues of environment-behavior studies and applications.[5] In
Europe, a series of biennial conferences has been held since 1970
under the title of architectural psychology, and has resulted in the
formation in 1981 of the International Association for the Study of
People and Their Physical Surroundings (IAPS) and the publication
of the *Journal of Environmental Psychology*.

The rapid expansion of environment-behavior studies and environmental design research is reflected also in the appearance of numerous textbooks and readers that comprehensively survey the field,
the establishment of several journals (*Environment and Behavior*,
1969; *Population and Environment*, 1978; *Architecture et Comporte-
ment/Architecture and Behavior*, 1980; and the *Journal of Environmental Psychology*, 1981), the publication of six monograph series
focusing on theoretical, empirical, and applied advances (for example, the Advances in Environmental Psychology Series from Erlbaum,
Environmental Design Series from Van Nostrand Reinhold, Environment and Behavior Series from Cambridge, Human Behavior and Environment Series from Plenum, and Environmental Psychology
Series from Praeger). Moreover, the increasingly international scope
of the field is evident from the recent professional meetings that have
been held in Australia, Canada, England, France, Germany, Japan,
the Netherlands, Scotland, Spain, Turkey, the Soviet Union, as well as
many in the United States, and the establishment of graduate training
programs in the social sciences and the environmental professions
within many of these countries. Judging from these developments,
environment-behavior studies and environmental design research
seem to have established themselves as a viable combined field of
theoretical and empirical research and research applications to environmental problems.

As a result of all of this activity, the field has developed coherent statements of definition and purpose that are agreed upon by members originating from different parent disciplines: Zeisel from sociology (1975); Stokols from psychology, (1977, 1981); Villecco and Brill from environmental design and architecture (1981). Previously characterized as being at a conceptual crossroads in the 1980s (Moore, 1980b), it is time now for the field to develop a framework for identifying the major research areas and problems needing attention and an agenda for directed research and action. It is thus a major objective of this book to establish this framework, to identify research areas and problems needing immediate attention, and to suggest important research directions for the future.

THE UTILITY OF ENVIRONMENTAL DESIGN RESEARCH

Environmental design research is an emerging discipline. It is youthful enough to provide discussion and thought, but it is nevertheless old enough to have produced results and impacts.

Based on research done in the 1960s at the National Bureau of Standards and a current major study of human performance in office buildings (see interview by Gaskie, 1980), studies indicate that the ratio of personnel to capital costs in major corporations is about 30:1 (Villecco & Brill, 1981).[6] First year costs are in the range of 2:1 between salaries and construction, but over an expected 25-year life span of the building, capital costs drop to less than three percent of the total operating costs of large corporations. It is clear from these figures (see Table 1), that if there is a causal relationship between the design of the office environment and the productivity of office workers, the economics of behaviorally supportive design are very attractive (Villecco & Brill, 1981).

This relationship has been demonstrated. Numerous studies have shown that satisfaction with the work environment is a component of overall job satisfaction, and that job satisfaction affects the productivity of organizations by lowering absentee rates, turnover, and grievance actions (Sundstrom, Kastenbaum, & Konar-Goldband, 1978; Villecco & Brill, 1981; Brill, 1982). Surveys in England (Langdon, 1966), Sweden (Lunden, 1972), and by Louis Harris and Associates in the United States have shown that the office environment is one of several factors very important for job satisfaction (41 percent of a national sample: Harris, 1978). More direct linkages between job performance and environmental qualities, such as privacy, lighting, and furniture systems have been found (Villecco & Brill, 1981).

In a critical review, Brill (1982) concluded:

> If the amount of quality of major cost items in the office, such as furniture, privacy, speech privacy and noise control, and amount of space, were *all* increased for each office worker, it appears that substantial investment could be made in more behaviorally supportive design and still be cost-effective in increasing productivity and the quality of work life for all office workers (p. 50).

The studies that led to these findings have been conducted for U.S. and overseas corporations. As a result, new guidelines are being applied to redesigning existing office space as well as planning new offices.

One of the early works to have a major impact on environmental policy was the study by Schorr, *Slums and Social Insecurity* (1963). This study probed into housing from various angles to see how it relates to the lives and the patterns of adjustment of poor people. Based on evidence from the social sciences, Schorr showed that housing affects perception of one's self, contributes to or relieves stress, and affects health.

While not alone in making these claims, Schorr's book was influential in government circles. He found a wide range of architectural, political, and social decisions about housing that have explicit consequences for social security versus social insecurity. Highlighted were elements of housing that have the most impact on poor people: (1) the design and adequacy of the house or apartment itself; (2) the physical structure of the neighborhood; and (3) the amount and character of planned open space.

Table 1: Percentages of Total Cost of Achieving the Mission of the Office (calculated in constant $)

Mission Cost Component	*Time Frames for Costs in Years*		
	1 year	*10 years*	*25 years*
Construction, furnishings & equipment	38.1	5.8	2.8
Operations & maintenance	2.2	3.4	3.6
Office workers' salaries	59.7	90.8	93.6

Note: Assumptions for the table include the following: Space per worker: 165 sq. ft., including support space. Source BOMA 1976-79 reports and Canadian Department of Public Worlds National Survey; construction cost: $50/sq. ft., National average January 1980; furnishings: $1,500/worker, estimated; complete replacement after 10 years; energy costs: $1.02/sq. ft., increasing at 15%/year; maintenance and operating costs: (excluding energy) $2.53/sq. ft., increasing at 8%/year. Source: BOMA; salaries: Assume engineering technician, grade 4 at $15,221.

Source: Reprinted with permission from Villecco and Brill (1981).

Having identified the elements of housing that affect poor people, Shorr then turned to considerations of economics and politics in order to change the situation facing the poor. The study concluded with a series of policy recommendations for both public and private sectors.

Schorr's appraisal of the effectiveness of housing policies in the United States was conducted from within the Social Security Administration of the former U.S. Department of Health, Education, and Welfare, with assistance from the Housing and Home Finance Agency and the Federal Housing Administration. In the mid sixties, Shorr's work led, in part, to the formation of the U.S. Department of Housing and Urban Development, specifically chartered to give cabinet status to efforts to confront poverty and to provide housing, urban neighborhood, and open space development for all Americans.

Other research on environment-behavior issues has had significant impact on policy, planning, and design. Vandalism and crime have been major and increasing problems in public schools, housing, public transportation systems, recreational facilities, and public spaces. The expense of property damage through vandalism in public schools cost the United States about $220 million in 1975 (National School Public Relations Association, 1975) and an estimated $377 million in 1981 (Brill, 1982).

Environmental design research looked at this issue. The findings of two studies indicate quite clearly which environmental design factors are related to a low incidence of school vandalism: aesthetic quality, good maintenance, location of schools in areas of diversified usage and high activity levels, natural surveillance, and location in areas of high illumination (Pablant & Baxter, 1975; Allen, 1978). These studies led to a series of design guidelines for reducing school property vandalism (see, for example, Zeisel, 1976) published by Educational Facilities Laboratories and the American Association of School Administrators for implementation in the nation's schools.

In public housing, though the dollar value of vandalism has not been calculated, Newman (1976) has described vandalism as a prime factor in the cycle of disintegration of public housing in the United States:

> High crime and vandalism are making these projects unlivable, even for families who have little choice in housing. Once the process of community disintegration has gotten underway, it is almost impossible to reverse. New families cannot be enticed to move into these developments and existing families only wait for

an opportunity to move out. Vacant units are vandalized to the point which they cannot be rehabilitated easily, and criminals, vagrants, and drug addicts use vacant units as a base of penetration against residents.

Newman's studies (published widely in *Newsweek, Time, Intellectual Digest,* and the New York *Times*) uncovered four major factors that contribute to the reduction of vandalism in housing: natural surveillance, or what Jacobs (1961) earlier called "eyes on the street"; well-defined territories; an image of security and upkeep; and proximity to "safe zones," to areas or buildings seen as safe, like religious buildings or highly used commercial centers (see Newman, 1972, 1973, 1976, 1980). Many housing projects around the country have been directly affected by these findings. The "Crime Prevention through Environmental Design Program" developed by the U.S. Law Enforcement Assistance Administration is another example of the impact of this research on design standards.

Environmental design research input to public policy and design programs has been documented in other places. The American Psychological Association Task Force on Environment and Behavior examined various mechanisms by which research on environment and behavior has served as viable inputs to public policy and design programs at the national, state, and local levels (Archea & Margulis, 1979). The Task Force found that the greatest emphasis in the late 1970s was input to regulatory, legislative, and adjudicatory decisions bearing directly on actions of designers and managers of the environment.

One of the most dramatic cases examined involved the development of a preparation program to reduce the consequences of placing elderly people in nursing homes. Environment and behavior research (Bourestom & Tars, 1974; Bourestom & Pastalan, 1975) documented a dramatic increase in the mortality rate of elderly persons who were forced to move from one institutional setting to another. These increases appear to be directly related to a combination of physical decline and environmental changes. Reports of the research were made available, and an experimental relocation program aimed at attenuating the fatal effects of "transfer trauma" was begun in Michigan. The power of the research in affecting policy, regulatory, and judiciary decisions can be seen by the following: (1) in the three-year period, seven civil rights cases in five states used these findings as a basis for court decisions; (2) guidelines based on this research have been used by public interest lawyers and by social workers involved in such cases; (3) regulations to protect the institutionalized elderly were adopted by at least one state; (4) memoranda urging all states to adopt similar programs were issued by two federal agencies; (5) 11 bills requiring a preparation program were introduced in the United States Congress; (6) draft preparation guidelines were issued to the states by the U.S. Administration on Aging; (7) a clearinghouse service used by lawyers published a widely read article on transfer trauma; and (8) the former U.S. Department of Health, Education, and Welfare promulgated the regulations as part of the enforcement of Public Law 92-603 on the environmental and medical aspects of nursing homes and other facilities for the care of the dependent elderly (Archea & Margulis, 1979).

In recent years, increasing concern for the environment aspects of social issues such as crime, imprisonment, aging, consumer safety, and fire protection has been expressed. The U.S. Consumer Protection Safety Commission has identified both stairs and children's playgrounds as being among the top ten causes of consumer accidents. In both cases, the National Bureau of Standards has been asked to conduct environment-behavior research on the causes of accidents, and to recommend new planning and design standards. As an example, the results of research on stair safety indicate a relationship between accidents and the riser/tread relationship with 11 inches being recommended as the minimum tread depth (Archea, 1977).

For playgrounds, the single greatest cause of accidents has been found to be the type of surface under play equipment. A staggering 72 percent of all injuries are caused by falls directly to a hard surface (U.S. Consumer Product Safety Commission, 1975). Studies at the National Bureau of Standards have made it clear that the safest materials are loose materials (Mahajan & Beine, 1978), and that the depth of these materials should be approximately the same number of inches that the maximum possible fall could be in feet (for example, a minimum of 4 inches for a maximum 4-foot fall; Moore, 1982b). These findings have been incorporated into a new set of national recommendations for playground safety issued by the U.S. Consumer Product Safety Commission (1981) that are directly impacting playgrounds across the country and cases pending in the courts.

One of the most powerful demonstrations of the impacts of environment-behavior research on new planning and design standards is the revision of the American National Standards Institute (ANSI) voluntary standards on access to buildings for the visually and physically handicapped. A state-of-the-art literature search was followed by field and laboratory research and the simulation of potential economic consequences of various regulatory changes (Steinfeld, 1979, 7 volumes).

In 1981 and 1982, environmental researchers evaluated the nature of the technical knowledge, most of it from published sources, underlying the exit facility design and emergency escape provisions of the National Fire Protection Association Life Safety Code (1976 edition). This code is probably the single most widely referenced guide to building circulation design, particularly for places of public assembly. A report prepared at the National Bureau of Standards (Stahl, Margulis, & Ventre, 1982) was used by the committees drafting the mandatory escape provisions for the revised edition of the code. An earlier National Bureau of Standards report (Ventre, Stahl, & Turner, 1982) identified successful practices, rather than published research findings, and incorporated these into voluntary guidelines for designers and managers of places of public assembly (see also Ventre, 1975, 1982).

Widely read news journals have remarked on the importance of environmental design research concerns in articles and editorials. The July 29, 1982 issue of *Engineering News Record* reported on a U.S. General Accounting Office report taking the U.S. General Services Administration to task for what the congressional watchdog agency believed was "inefficient transmission of energy-saving data to designers of federal buildings" (p. 13). In order to identify and alleviate problems like those discovered at the Norris Cotton Federal Office Building in Manchester, New Hampshire, the GAO report also recommended that GSA "implement a post-occupancy evaluation program" (p. 13). The *Wall Street Journal*, July 30, 1982, reported to its readers the findings of studies conducted at MIT that open spaces stimulate communication and increase productivity in office buildings (see also Wineman, 1982). The results were immediately applied in the design of a new $14.8 million engineering and research office building for Corning Glass Works in Corning, New York. The *Wall Street Journal* also reported that a growing number of companies are using the ideas of the environmental social sciences to design buildings, including American Telephone & Telegraph, Bell Telephone Laboratories, IBM Corporation, and Exxon Production Research, a unit of the Exxon Corporation. Being aware of such research and its uses, the American Institute of Architects passed a resolution at its 1982 convention to officially back the development of a "knowledge base" for all environmental planning and design decisions (AIA, 1982).[7]

These are a few examples of the utility of environmental design research. In summary, we have seen in this chapter that research in the field of environment and behavior has direct application to the problems facing society. A fundamental aspect of environmental design research is that it is applied as well as basic in nature. It has led to major impacts on various aspects of improving the quality of the physical environment. It has led to important national, state, and local policies. It is cost effective in that a small amount of money spent on environment and behavior research can have enormous multiplier effects (as much as 35 times, as in the office worker satisfaction and productivity studies). In the past decade, environment and behavior research has made important contributions to health, safety, and welfare, and to the enhancement of the quality of life.

NOTES

1. The complete history of the field is yet to be written. For partial histories, see Carson (1965), Archea (1975), Proshansky and O'Hanlon (1977), Stokols (1978), and Proshansky and Altman (1979). As pointed out by White (1979) in her introduction to the Proshansky and Altman review, the authors of these histories have chosen to focus their overviews on particular portions of the multidisciplinary field, and in particular to "environmental psychology," and not to the broader field of environmental design research (environment-behavior studies and applications). At the time of this writing, two additional histories are being prepared; see Moore (in press) and Bechtel (in preparation).

2. Roger Barker received the 1981 EDRA Career Award for his seminal influence on the field.

3. Another EDRA Career Award recipient, Amos Rapoport, was honored for these contributions in 1980.

4. Irwin Altman received the EDRA Career Award in 1982 for these and other important contributions to the field.

5. As of 1982, the following scholarly and professional associations had sections focused on environment-behavior research and/or applications: the Division of Population and Environmental Psychology of the American Psychological Association; the Environmental Sociology Section of the American Sociological Association; the Environmental Perception Specialty Group of the Association of American Geographers; the Design and Environment Section of the Gerontological Society; the Environmental Design Technical Interest Group of the Human Factors Society; the Environmental Psychology Section of the International Association of Applied Psychology; the Research Corporation of the American Institute of Architects; the Research Office of the American Planning Association; the Interior Architecture Section of the American Society of Interior Designers; and the Landscape Research Section of the American Society of Landscape Architects.

6. Much of the following discussion and examples of office buildings is from the research of Michael Brill and his associates (Villecco & Brill, 1981; Brill, 1982). Michael Brill received the EDRA Career Award in 1979 for these and other seminal contributions to research and application.

7. Our sincere thanks to Francis T. Ventre formerly of the National Bureau of Standards and the National Science Foundation for making many of these examples available for us.

3

GOALS, VALUES, ORIENTATIONS

The major goals and assumptions of the field, including its attention to improving the quality of life through the quality of the environment, its commitment to problem solving, its equal emphasis on all scales of environment, and on time, change, and adaptation, and its concerns for everyday experience and behavior, the integrity of person-environment events, and the real, physical environment of human experience.

Environmental design research is a *field* in the sense that there are certain assumptions, values, and points of view that guide and give it structure and direction. These values and orientations may provide the foundations for a new, theoretically coherent and practical discipline bridging the social and design sciences.

Environmental design research is value-explicit. Historically the view was that science was value-free, that the enterprise of science was the search for truth, wherever it might be found, and that personal or societal values should not play a role. Contemporary philosophers of science now acknowledge that science is not value-free, but rather that it should be value-explicit (e.g., Kuhn, 1962). In environmental design research, the decision of which environment-behavior interactions to study is fundamentally driven by individual and social values. Making a value check is inherent in the decision to study a certain impacted population, or a certain building type, or a certain socio-cultural phenomena. To study slums and social insecurity (Schorr, 1963), the determinants of neighborhood quality (Marans, 1979), or worker satisfaction in modern office buildings (Langdon, 1966) is a question of values. To do action research on a particular contemporary issue is clearly a value choice.

Since consensus and conflicting value systems determine decisions about research and interventions in the designed environment, these values must be made explicit. They are knowable, and they influence conceptual formulations and policy decisions. In environmental design research, the values underlying any project should be made explicit, open to scrutiny, debated, and weighed for their implications for future policy and practice (see Knight & Campbell, 1980).

A conflict may exist between long- and short-run benefits of research in the built environment, between the goals of the marketplace and those of optimizing the quality of environments in terms of human growth and well-being. These "conflicts" may serve as a healthy tension, if both directions are adequately supported.

Of ideological concern to the community of researchers is also the nature of the relationships between research and practice. This set of concerns addresses the divergent nature of objectives within most modern societies. Is the purpose of a prison to rehabilitate, to punish, or to isolate? Of a school to teach fundamentals or to create an atmosphere for socialization? Of a museum to instruct or to provide recreation? Of housing to shelter or enhance family and community relations? Of a park for individual and group satisfaction or for open space conservation?

Within this set of concerns lies the fundamental issue of who should be the primary client. To whom should environmental design research be addressed: the paying or the nonpaying client, or science itself, or society (Zeisel, 1974)? The issue of who the client is also encompasses varied beliefs of the role of the environmental design and planning professions.

This chapter briefly characterizes some of the values and orientations that guide the field.[1] They have been organized into five categories: goals; contextual values; conceptual orientations; methodological values; and future orientation.

GOALS OF ENVIRONMENTAL DESIGN RESEARCH

Improving the Quality of Life

A fundamental goal in the field is the search for the attributes of quality in the physical environment and their consequences for the quality of life.

The assessment of consequences is a fundamental mission of the social sciences. The assessment of the quality of life is also a priority in many professions and at the highest levels of government

(President's Commission for a National Agenda for the Eighties, 1980). That the quality of life and environmental quality are also becoming critical to the building sciences and the design and planning communities, provides a unifying cord between disciplines and between research and policy.

The nature of the search process and the elements that determine congruency between humans and the built environment are part of a continuing debate. They form the basis for varied programs of research. The criterion by which quality or qualities are defined remains a major source of philosophical difference among researchers and probably constitutes one of the major differences in perspective among designer-practitioners, managers of the designed environment, and researchers.[2] Viewing these differences as creative tensions rather than hard conflicts will produce an atmosphere for improved communications. Producing in the office environment, and learning in the educational environment are criteria that have a place in environmental design research. The extent to which these same criteria are convergent with the criteria that may emerge from environmental users themselves is an important question. In many societies and settings, individual control and reinforcement of collective values is a priority issue.

While the field has a value commitment toward improving the quality of life, Brill points out:

> Our research *could* be used in ways we would not approve of. We cannot control how it is used once it is in the public domain or even in the hands of our clients or sponsors. We cannot insert values in the research questions, procedures, or results which could control their "proper" use.[3]

CONTEXTUAL VALUES

Problem-Centered and Action-Oriented

A fundamental commitment to the quality of life and the consequences of the physical environment to same, leads directly to an orientation to research the problems that people experience. Through research that better informs policy, planning, design, and education, environmental problems must improve.

The commitment of environmental design researchers to a better world, and to the quality of life and the impact of the environment on that quality, may be termed a problem-centered focus. As pointed out by Proshansky and Altman (1979), environment-behavior theory and

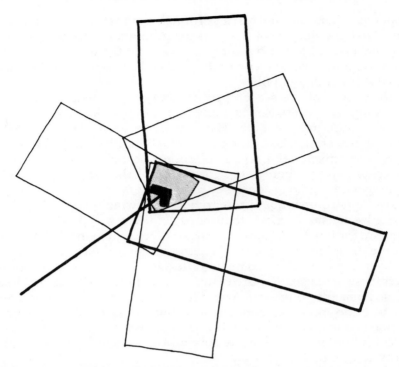

Figure 4 Environmental design research is problem-centered and action-oriented.

research begin and end with human-environment problems that characterize a society, at any scale of environment, and at any period of time. While current research focuses on fairly narrow issues of the psychology of the individual in social space (e.g., most issues of *Environment and Behavior* or the *Journal of Environmental Psychology*), it is a deep and abiding commitment of the field to sharpening awareness of environmental problems.

This orientation also leads to the field being deeply committed to communication between research and application, or the human sciences and the design professions. It also leads to explicit use of the theories, findings, and research strategies of the field in developing solutions to environmental problems, while communicating these solutions to a wide professional audience (e.g., the design guidance series of the U.S. Army Corps of Engineers).

Environmental design research may also be said to be action oriented. Large portions of the field are concerned with interventions. Those aspects of the field are committed to environmental policy formulation, environmental management, impact assessment, urban planning, architecture, and landscape architecture. While these portions are committed to interventions, and have to make tough decisions, they are distinguished from traditional aspects of the environmental professions in that they are always on the lookout for the best research information on which to base environmental decisions. This is what distinguishes an environment-behavior orientation to architecture and the allied professions, from a more formal approach to design.

Since research is value-explicit, conducting research for the advancement of knowledge alone is not enough. The field must carry forth knowledge into research utilization.

All Scales of the Physical Environment

From the shape of whole regions down to the design of window hardware, environmental design research gives equal emphasis to the growing concern for resource conservation and productivity while addressing the wide range of scale of design efforts to deal with these problems. It is apparent that natural resource decisions, for example, can affect both the physical form of a region and the health and behavioral consequences for residents and visitors. Environmental design research involves all levels of analysis and all scales of the physical environment. Those most interested in micro-environments (e.g., in human factors or in interior design) look at issues of person-environment fit at the room scale or smaller (e.g., Dreyfuss', 1966, research and charts on anthropometrics, or Helmreich's, 1974, studies of undersea habitats). Those most interested in meso-scale environments (e.g., in environmental psychology or in architecture or landscape architecture) tend to look at the building or neighborhood scale (e.g., Hershberger's, 1968, studies of meaning in architecture, or Farbstein & Wener's, 1982, studies of correctional facilities). Those most interested in macro-scale environments (e.g., in sociology, geography, or in urban planning) tend to look at the city or regional scale (e.g., Michelson's, 1977, research on housing choices, or Zube's, 1976, work on landscape and natural resource perception). All scales are a part of environmental design research. As stated in Chapter 1, the *design* in environmental design research means the conscious effort to influence the physical environment at all scales and in all time frames for the improvement of the quality of life.

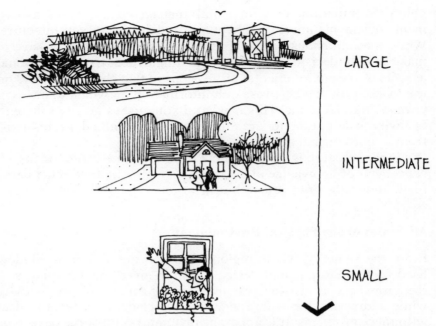

Figure 5 Environmental design research deals with all scales of the physical environment from the space of objects to the space of a region.

Time, Change, and Adaptation

Events occur in space and in time. Changes occur across space and across time. Environment-behavior relationships occur and extend over short and long periods of time. They have their origins in historical time. Concepts of change and adaptation appear frequently in the literature. This focus on the dynamic interaction between human behavior and both the scale and shape of designed environments reflects a growing rejection of the more simple cause-effect paradigms of traditional social science disciplines.

Time is one of the explicit focuses of recent studies in the field (e.g., Lynch, 1973; Wapner, 1981). It is also an issue that raises questions: Are we studying behavior and choices that are a result of past influences? Can we predict what people will prefer and how they will react in innovative environments? The investigation of time in environmental design research has at least these two components: the study of environment-behavior relations across time; and the prediction of the future impacts of innovative environmental interventions (design, planning, and policy, e.g., as characteristics of user groups change over time).

While sympathetic with decision makers' "need to know," environmental design research believes its exploratory mission extends beyond project inquiry toward better understanding of both behavioral and cultural issues and a better grasp of key design and spatial definitions. The products of varied design research efforts are working process documents to be replicated and retested, not final statements of effects to be purposely set into program documents. Research into the transactions between people and built environments is viewed as an *iterative process* with products at each loop and with longer time spans than those processes that constitute a design-build sequence.

Multidisciplinary by Nature and Necessity

The problems of the human environment are rooted in complexity. At each level, situational, historical, and cultural factors condition the interplay. To understand this complexity, environmental design research seeks multi-disciplinary collaboration in its research orientation and in environmental applications.

This linkage across divergent disciplines must go beyond simple communication and joint effort and begin to develop a language rooted in a newly conceived integrative theory of environment-behavior-design relations (see, for example, Stokols, 1977a; Proshansky & Altman, 1979; Villecco & Brill, 1981; Zeisel, 1981). To date, there is no one integrative theory rising above all others. This is not a bad state of affairs for an emergent field at the end of its second decade. Rather, support is required for fundamental investigations that have potential to contribute to the conceptualization and test of major integrative theories.

CONCEPTUAL ORIENTATIONS

The Everyday Physical Environment of Human Experience

A fundamental belief in environmental design research is the importance of studying human experience and action in the everyday physical environment (Craik, 1968). Real life settings are paramount in studies. Behavior, experience, and action are seen in these contexts, rather than studying isolated phenomena relative to isolated variables. The orientation of environmental design research is to study people in groups as they carry out their normal activities (Gans, 1959).

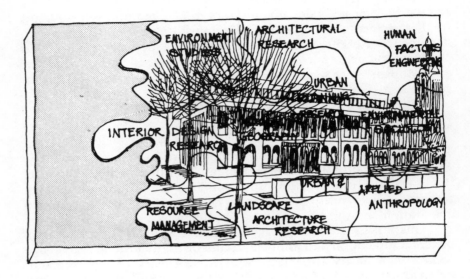

Figure 6 Environmental design research is multidisciplinary.

The Integrity of Person-Environment Events

Environmental design research resists thinking of people or settings independently of each other, or thinking of research and application separate from each other. Events in the world involve characteristics of people and of the settings in which they are embedded. It may even be said that environment-and-behavior is a transactional unity—a single unit of analysis.

In the seminal works of Lewin (1936) and Barker (1968), behavior in everyday settings is seen as a joint product of human forces and situations factors. While some argue that the field is comprised of three axes: people, space, and culture (e.g., Rapoport, 1976), others suggest that events are affected by internal or organismic factors, including intrapersonal and interpersonal processes of individuals, groups, and cultures, and situational or

Figure 7 Environmental design research respects the integrity of person–environment events. Events are the product of internal organismic forces and environmental factors (after Piaget, *The Development of Thought*, 1975).

environmental factors, including the social, cultural, and physical environment (Moore, 1976; Stokols, 1977a).

Content as Much as Process[4]

Many social sciences, especially psychology, have focused primarily on processes—processes of human development, processes of cognition, and processes of group dynamics. From an examination of the literature in environmental design research (see, for examples, the annual EDRA *Proceedings*), it is evident that far more attention is given in this field to the *content* of these relationships—who the actors are, what activities they are engaged in to achieve what purposes, and in what physical settings. A subtle distinction can be

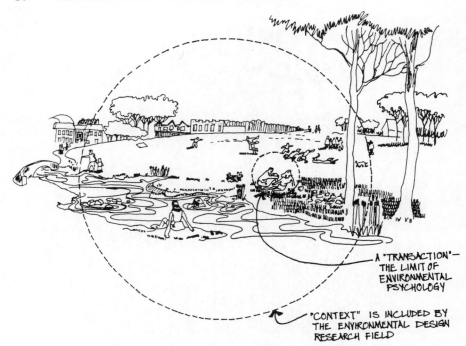

A "TRANSACTION"—
THE LIMIT OF
ENVIRONMENTAL
PSYCHOLOGY

"CONTEXT" IS INCLUDED BY
THE ENVIRONMENTAL DESIGN
RESEARCH FIELD

Figure 8 Environmental design research investigates content as much as process.

made here between environmental psychology and environmental design research. In a recent review (Russell & Ward, 1982), environmental psychology was characterized as focusing upon intrapersonal processes, such as perception, cognition, and learning, that mediate the impact of the environment on the individual. The more encompassing field of environmental design or environment-behavior research includes these aspects, but also includes the broader questions of group behavior, social values, and cultural norms in relation to the environment.

The Mediating Role of Psychological, Social, and Cultural Processes

While stating that environmental design research is concerned with the content of environment-behavior relations, the field also takes the position that the relations between environment and behavior are mediated by processes that are internal to the person, the group, and the culture. These are seen as mediating or intervening variables. In environmental design research, the environment is not

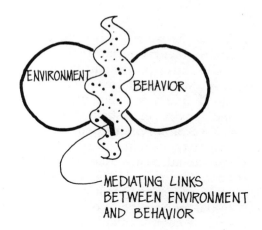

Figure 9 Environmental design research investigates the mediating links between environment and behavior.

seen as having a direct impact on people and groups; rather it has an impact via people's perceptions and conceptions of it (Moore & Golledge, 1976). Environmental design research is eclectic, with different scholars and researchers using various intervening constructs to interpret, explain, and predict environment-behavior relations (Michelson, 1976).

METHODOLOGICAL VALUES AND ORIENTATIONS

Descriptive, Exploratory, and Quasi-Experimental Research

Environmental design research has several strong methodological orientations. These orientations are derived from the nature of the phenomena that are studied in environmental design research, and the nature of the applications of this knowledge to policy, planning, design, and education contexts.

First, the description of environment-behavior phenomena must be the primary element of environmental design research. As this field is new, there are few stable descriptions of phenomena to fall back on. The past 20 years have seen the beginning of efforts to record descriptive properties and relationships of environment-behavior situations relative to this field. They must continue to be established by observation before we can dare to try to explain them, or to predict them. It is also necessary to know what environmental and personal variables might be operating in an environment-behavior situation before trying to do parametric studies of the causal relations between independent and dependent variables. Thus, for environmental design research the basic methodological tools are qualitative and quantitative accounts of people-place transactions in the form of case histories, phenomenological accounts, open-ended interviews, longitudinal observations, and visual records. Once environment-behavior phenomena have been described and key variables identified, more traditional social science quantitative methods come into focus.

Second, environmental design research is characterized by exploratory research, not necessarily research to test specific hypotheses derived from general theories. Exploratory research implies an interest in the molar aspects of the environment and a willingness to deal with the complexity of phenomena in their everyday terms. This orientation follows from the field's interest in the content of environment-behavior transactions. Early contributors to the field adopted a molar and exploratory stance that was in contrast to the prevalent values of their parent disciplines (e.g., Sommer, 1969; Proshansky, 1972). Looking at a phenomenon for what it is, describing the phenomenon, developing a longitudinal case history of it, speak well for exploratory research. It appears that the field is now ready to move beyond these exploratory beginnings to the development and testing of alternative major theories (see Chapter 5).

Third, there is considerable interest in the field in conducting rigorous studies with appropriate concern for internal and external validity, but still in field settings. The methodological advances of quasi-experimental design in field settings (Campbell & Stanley, 1963; Cook & Campbell, 1976; and Cook & Campbell, 1979) are being translated and applied extensively in environmental design research (see, for example, Patterson, 1977).

FUTURE ORIENTATIONS

Futures

It is noteworthy that while a majority of the activity in the field is more oriented to current events, a significant thread of future orientation necessarily enters into discussions of design and of environmental research needs (e.g., at the annual EDRA meetings). Expressions of concerns for preservation of natural and historical settings are balanced by sharp realization of emerging technology and styles of living and working. Future-oriented discussions in the field encompass the development of new tools for design as well as advanced and complex methods of research and research dissemination.

Inevitably, discussions of environmental futures involve attitudes and beliefs about the role of environmental design research in relation to public policy, and an awareness by the research community that, on the one hand, choice of research priorities may be formalized by prevailing policy and, on the other hand, research may lead to policy confrontation and contradiction.

NOTES

1. This section is based on responses to the EDRA Survey on Research Need and Directions. For additional discussion on some of these assumptions, see Proshansky and Altman (1979), Stokols (1977a), Wohlwill (1970), and Moore and Golledge (1976).

2. "Designers and aestheticians argue about how 'quality' is defined (monism); scientists and humanists argue about how 'qualities' are defined (pluralism)." F.T. Ventre. Personal letter communication, December 23, 1981.

3. M. Brill. Personal letter communication, January 8-11, 1982.

4. This section is based on Proshansky and Altman (1979) who first made clear the distinction between the process orientation of the social sciences and the content-plus-process orientation of environmental design research.

4

AN ORGANIZING FRAMEWORK

A conceptual framework for synthesizing the many contributions of the field over the past 25 years, and showing the gaps and needed research directions for the next decade. Includes the notions that all issues of environmental design research or application to professional practice necessarily involve considerations of places, environmental user groups, socio-behavioral events, and time; that the role of theory is to explicate the relations between these factors; that environmental design research and application is a cyclical process; and that there is a larger context of cultural and environmental problems that simultaneously impel the field onward and constrain its development.

To show the interrelationships among the various parts of environmental design research and the relationship to other areas of scientific research and professional practice, a framework for environmental design research is presented in this chapter. Earlier frameworks have been developed by Craik (1968), Altman (1973b), Moore (1979a), and Villecco and Brill (1981). The framework presented here extends those earlier statements.

In any field, a framework describes the conceptual organization or structure of the field. The purpose is to help people in the field relate to each other in a consistent way, and to help those outside the field understand it. Frameworks are not theories. They do not attempt to explain the phenomena discovered in the field. But they are more than orienting assumptions, as they also highlight the structural connections between different parts of the field. Like theories, frameworks are not right or wrong, just more or less useful to the degree that they present a structure that is useful to the field. And, like theories, they are evolutionary, building on earlier statements and setting the stage for their own demise as a better, more inclusive, and more useful framework is described in the literature of the field.

The framework presented here consists of four components:[1] (1) the notion that any environmental design research question is inherently defined in terms of place, environmental user group, socio-behavioral events, and time; (2) the proposition that the role of theory is to elucidate the relations between these dimensions; (3) the process of iterative environmental design research and applications; and (4) a context of cultural and environmental factors acting on the field. The chapters that follow in Part II emerge from this conceptual framework.

PLACES, USER GROUPS, SOCIO-BEHAVIORAL PHENOMENA, AND TIME

The first component of the framework is the notion that any environmental design research question can be seen along four dimensions (see Figure 10): places, people, socio-behavioral phenomena, and time (Altman, 1973b; Moore, 1979a; Zeisel, 1981; Stokols, 1981b). This is, and always has been, the crux of the field.

Environmental design research deals with particular types of environmental settings such as housing and factories. Particular attention should be paid to groups of environmental users which might include children, factory workers, and those in health-care facilities, and those like the elderly and the handicapped who are specifically influenced by their interaction with the environment.

Environmental design research is also concerned with the dynamic interaction between the three dimensions over time. Environmental design research deals with events that change over time. The concepts of change and adaptation are critical concepts in the field, both change in human expectations and adaptation of the environment to human changes.

Any one of these four dimensions may be the prime focus for a program of research. Thus, research on crowding or environmental cognition focuses on the socio-behavioral dimension (e.g., Stokols, 1972a,b; Moore & Golledge, 1976). Research on housing or environmental resources focuses on place types (e.g., Michelson, 1977b; Craik & Zube, 1976). Research on children or the handicapped are environmental user group issues (e.g., Howell, 1980; Weinstein, 1979; Steinfeld, 1979). Investigations of environmental adaptation and coming to terms with a new environment are examples of research focused on the time dimension (e.g., Wapner, 1981).

THEORY

The second major component of the framework is the development

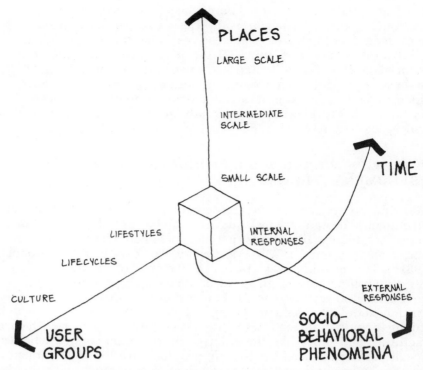

Figure 10 Four dimensions of analysis in environmental design research: places, user groups, socio-behavioral phenomena, and time (a variation on Altman, 1973b, who identified environments, behavior, and the processes of application, and on Moore, 1979a, who suggested places, people, and phenomena).

and use of integrative theory. The dynamic mechanisms that link the dimensions of people, place, behavior, and time are explained through the medium of theory. Theory is the development of a systematic set of assumptions, accepted principles, and constructs devised to explain the nature of a specified set of environment-behavior phenomena. The development and test of integrative theory of environment-behavior relations is one of the most pressing needs for the immediate future.

PROCESS: THE FEEDBACK BETWEEN RESEARCH, EVALUATION, AND APPLICATIONS

The third aspect of this framework is the notion that design and

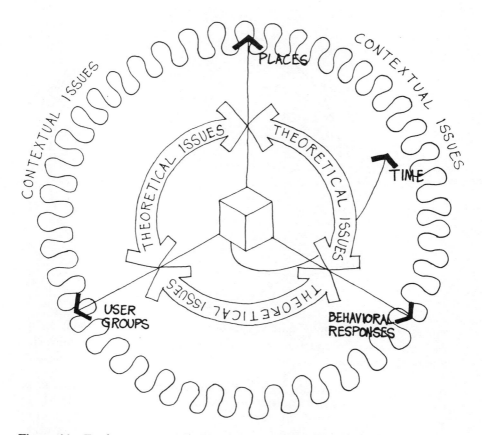

Figure 11 Explanation of the dynamic relations among the four dimensions.

research, while contributing to each other, are distinct conceptual entities, but are tied together in an iterative process (Zeisel, 1975, 1981; Villecco & Brill, 1981). Research involves the systematic analysis of phenomena under conditions allowing facts, laws, and theories to arise. "Design" used in the broadest sense is the application of the knowledge gained to the solution of real-world problems in the everyday physical environment. Design practice and scientific research are distinct entities, each with its own systems and rules, but potentially sharing a common commitment to a body of knowledge (see Zeisel, 1981).

Virtually any environment-behavior topic can be researched that relates people to designed environments and changes in the

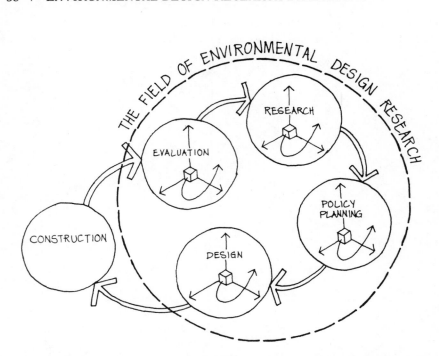

Figure 12 The cyclical and iterative nature of environmental design research—research and application of knowledge to environmental policy, planning, design, and education.

quality of life as a result. Traditional design practice, planning practice, and the process of public policy decision making frequently involve informal applications of knowledge, information, and research results. The products of environmental design research influence environmental policy, models of planned environments, and designs for particular buildings, landscapes, and urban places.

While research and design are conceptually distinct, they are not isolated from each other in the field of environmental design research.

The interaction between environmental design and research is cyclical and iterative (Zeisel, 1975, 1981; Villecco & Brill, 1981). Much of the field is applied in nature, that is, asking questions of immediate interest to environmental problem solving. Public policy formulation, environmental planning, facility programming, and post–occupancy evaluation are the places in the policy making/planning/design/build cycle where environmental design research has had the grestest impact. Wener (1982) has documented some examples of this process, including evaluations of environments designed in response to previous post–occupancy evaluations.

CONTEXT: URGENT EVIRONMENTAL
PROBLEMS AND CONSTRAINTS

Surrounding these four components, and especially the dimensions of places, people, and phenomena, are a number of contextual factors. They are examined in Chapter 10, and involve issues of environmental deterioration, depletion of natural resources, and improving the quality of life. These contextual issues set the agenda for the field, as they define the most critical issues facing our sociophysical environment and therefore facing our field. Environmental design research is devoted to the creation of knowledge on which to base public and private actions for resolving these important issues. The body of the report—Chapters 5 through 9—concentrates on needed areas of environmental design research as a means for expanding this knowledge base.

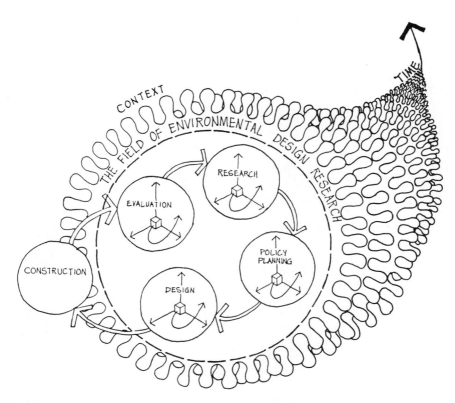

Figure 13   The total framework of environmental design research.

NOTE

1. This framework is a reflection of the content analysis of the 178 items submitted by respondents to the EDRA Survey, workshops, position papers, reviews, and other research agenda (see Appendixes A, B, and C), and is based on an earlier conceptualization by the senior author (Moore, 1979a) and by other writers in the field (most especially Altman, 1973b).

PART II
CRITICAL RESEARCH
ISSUES

5

THEORETICAL AND CONCEPTUAL ISSUES

The major conceptual issues facing the field. The values of theory, some dimensions along which environmental design theory can be developed, and characteristics of good theory in environmental design research. Distinctions between theoretical orientations, frameworks, conceptual models, and explanatory theories. Examples of the major theories in this field are given by way of illustration.

Fields advance in part by research focused on the development and testing of theory and by research exploring basic conceptual issues. In a problem-focused field, there is usually a continuum of contributors—from practitioners tackling intervention problems to methodologists addressing process issues, scientists investigating substantive issues, and theoreticians developing models and theories. If one of these groups unduly dominates a field, the field as a whole may not be adequately developed.

In environmental design research, sub-areas and different contributors have chosen to focus on various issues: theory, empirical or scholarly research, applied research, research applications, and environmental problem solving. This diversity of approaches in the field is positive, and should be encouraged. Environmental design research involves the spectrum from theory development to environmental problem solving.

It is important for the field of environmental design research to conceptualize those aspects of the environment and social behavior that offer the greatest promise of illuminating a general theory of environment-behavior relations (Chase, 1973). This chapter presents some priority issues in the development of theories of environment-behavior relations and highlights a number of pressing conceptual issues needing attention.

The discussion of theories will cover three basic issues: the value of theory; some dimensions along which environmental design theories can be developed; and nine characteristics of good theory in environmental design research. Examples of current theories are given. This will be followed by a discussion highlighting a number of other conceptual issues, the resolution of which may aid in the construction of an integrative theory for the field.

THE IMPORTANCE OF ENVIRONMENT-BEHAVIOR THEORY

The development and test of theories that relate environments and humans at all scales and level, with implications for "design" in the broadest sense, is a key aspect of the field. It deserves increased and sustained attention.

It is not surprising that calls for the development of theory did not occur during the first years of environmental design research nor the first few years of EDRA. At the third annual meeting of EDRA (EDRA 3, Los Angeles, 1972), a symposium was devoted specifically to theoretical issues in one area of environmental design research—theories of human cognition of the environment (Moore, 1972). The following year (EDRA 4, Blacksburg, Virginia, 1973), there were two sessions devoted to the comparison of different theories in the field in general (Altman, 1973a; Chase, 1973). Since that time, a number of advances have been made in the analysis of the general character of theory in environmental design research (e.g., Altman, 1973b; Rapoport, 1973; Archea, 1975).

The development and test of environment-behavior (or environmental design) theory is important for several reasons. Briefly summarized, there are six reasons (see also Rapoport, 1973; Patterson, 1977):

1. The development and articulation of major theories may signal the next development of the field of environment-behavior studies. Theory, research methods, and data cannot be separated; each ideally develops hand-in-hand with the others. It may also be that research applications cannot be separated from the other three.

2. The development of theory will help make sense of disparate data, and reveal gaps and inconsistencies in existing data as well as areas of agreement. While there are debates about the quality of data in the field (Danford, 1982; Moore, 1982a), environmental design research is relatively data rich. The two main research publications—the annual Environment Design Research Association *Proceedings* and the journal *Environment and Behavior*—are both in their second decades. While several thousand

scientific papers have been published in those pages, relatively few are concerned with the theoretical organization of large portions of that data.

3. The development of theory can lead the field from description to the explanaton of underlying processes. This is needed to give coherence to the field and its applications. At present the field is strong in description, but weak in explanation. One reason for this is the large-scale, long-term financial support necessary to develop programs of research that can integrate theoretical and practical problems.

4. An important role of theory is as a heuristic to suggest new lines of inquiry, new conceptual approaches, and new ways of conceptualizing new research and design problems.

5. Theory can aid in the application of findings to environmental problem solving—to help environmental professionals apply knowledge now available to aid in solving environmental problems.

6. Theory can aid in teaching environment-behavior studies and environmental design research, especially at the graduate level. It is much more effective to teach major theories, than it is to teach a myraid of unorganized findings.

The field of environmental design research is in a preparidigmatic stage (Kuhn, 1962). There is no single unifying theoretical perspective for the study of environment-behavior relationships. The study of behavior in designed settings proceeded from a ecological perspective, borrowing parts of the theories of Lewin (1946,

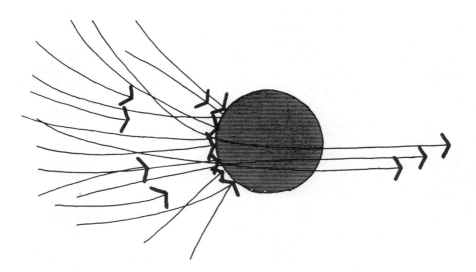

Figure 14 The need for theory in environmental design research.

1951) and Barker (1968). While this perspective is shared by many, it has not been embraced nor has any other theoretical perspective as a world view, paradigm, or agreed-upon explantory theory for the field.

Environmental design research is far from a state of agreement on theoretical orientation. The conceptions of our forerunners have seldom been built upon, tested, extended, or refined. As Merton (1957) has pointed out for sociology, theories are typically laid out as alternative and competing conceptions rather than consolidated into a cumulative product. There are significant exceptions to this statement (notably, the work of Barker's colleagues and students to develop and extend his formulations; cf. Bechtel, 1977; Wicker, 1979; Willems, 1977). A priority issue for the field, therefore, is the development, test, extension, and refinement of major unifying theoretical systems for environmental design research.

DIMENSIONS OF THEORETICAL DEVELOPMENT

There are significant differences in the use of word "theory" in science. Sometimes it is used to describe general orientations to research. Other times it is used to characterize a testable constellation of concepts, variables, and mechanisms used for the interpretation of particular phenomena.

It is possible to distinguish between four levels of theory of environment-behavior-design relations:[1] theoretical orientations, frameworks, conceptual models, and explanatory theories. Typically the development of theory proceeds through these four levels.

Theoretical orientations are broad conceptual approaches to a subject matter. They are heuristics that orient an investigator to look at environment-behavior phenomena in different ways and to identify interesting lines of investigation. Two well-known examples in environmental design research are the cultural orientation of Rapoport (e.g., 1969, 1976, 1979) and the personality orientation of Craik (e.g., 1970a, 1970b, 1976). The latter has led to the conceptualization and measurement of environmental dispositions and the utilization of established personality inventories to predict people's modifications of the physical environment, as well as its reciprocal impact on them (see, for example, Craik, 1970a; McKechnie, 1974; Bunting & Semple,

1979). While some theoretical orientations have led to tested propositions, a large part of what is now called environmental design theory consists of general orientations. "We have," to quote the sociologist Merton, "many concepts but few confirmed theories; many points of view, but few theorems" (1957, p. 9).

Frameworks describe the relations among existing findings in a particular area. A framework goes beyond an orientation in that it provides a systematic organization to data about different ways people and environments interact. An example of a systematic framework in environmental design research is Craik's (1968, 1970) characterization of data on environmental assessment. He organized the data in terms of: observers, environmental displays, response formats, and media of presentation. Chapter 3 of this volume is an attempt to provide an updated framework for the field of environmental design research.

Conceptual models are used to describe the dynamic relations of how people and environments interact. These models are quite similar to models in architecture and planning. In architecture, a model is often a static, iconic representation of some portion of the real world. Architects also use dynamic, symbolic models, as in the case of computer models of energy use in buildings. In planning, a model is often a dynamic simulation of events in the real world. It is based on descriptions of variables and incorporates statements about the presumed dynamic relations among variables. Models may be used to predict future events given certain parameters; they do not necessarily explain those events in a larger theoretical system. An example of a conceptual model in environmental design research is Marans' (1976) model of residential environmental quality.

Explanatory theories are systematic and testable constellations, concepts, and mechanisms explaining aspects of behavior in relation to aspects of environments. These theories attempt to explain why a set of observable phenomena behaves in the way it does. There are many examples of explanatory theories in environmental design research: the environmental competency theory of aging and the environment (Nahemow & Lawton, 1973; Lawton, 1980) and the theories of cognitive representation of the spatial environment (Moore & Golledge, 1976; see Stokols, 1978) are two.

A distinction may be drawn between what have been called "little t" and "big T" theories (Downs, 1976). Big *T* theories are coherent, explicit, and are intended to account for a wide range of data across several substantive sub-areas of a field, for example, the theory of relativity, probability theory, or behaviorist theory. Little *t* theories are also coherent and explicit but do not attempt to stretch beyond the substantive area from which they are developed; they account for limited bodies of data, for example, adaptation-level theory or location theory.

"Theories of the middle range" proposed by Merton (1948, 1957) are intermediate to the minor working hypotheses and the all-inclusive master conceptual scheme. Theories of the middle range are used to derive large numbers of empirically observed uniformities of social behavior. Prime examples of the beginnings of big *T*, or grand theories, in environmental design research are the field theory of Lewin (1946, 1951), the behavior-setting theory of Barker (1968; Barker & Wright, 1955; Barker, 1978) and the contextual theory of Stokols (1981b; Stokols & Schumaker, 1981). Examples of middle range theories are those explaining portions of the data in the field, for example, those mentioned above for development, aging, and the environment, and Altman's (1975) theory of privacy and social interaction.

It has been said that environmental design research is in a preparadigmatic stage (Stokols, 1977a) and that there are the beginnings of multiple competing paradigms (Craik, 1977). Merton cautions against the overreliance on grand theories, which he terms "all inclusive speculations." He states:

> I assume that the search for a total system of sociological theory, in which all manner of observations promptly find their preordained place, has the same large challenge and the same small promise as those all-encompassing philosophical systems which have fallen into deserved disuse. There are some who talk as though they expect, here and now, formulation of *the* sociological theory adequate to encompass vast ranges of precisely observed details of social behavior and fruitful enough to direct the attention of thousands of research workers to pertinent problems of empirical research. This I take to be a premature and apocalyptic belief. We are not ready. The preparatory work has not been done. (1957, p. 16).

If true for sociology, Merton's statement is more applicable for environmental design research, a younger field than sociology. However, the field of environmental design research seems overly data-rich when compared to the lack of theories to account for and explain that data.

A real need in the field is for continuing theoretical research on the construction and test of theories of the middle range. Other needs might include the construction and test of larger structures to explain the general nature of interactions between human behavior at all levels (physiological, psychological, individual behavior, sociocultural behavior) and the physical environment at all scales. What is needed is a shift in emphasis from a collection of data on single

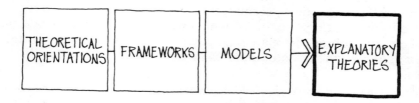

Figure 15   Dimensions along which theories can be characterized.

environment-behavior phenomena to the development and test of theories of the middle range, leading ultimately to the development of grand theories.

For the development of theory in the field, it is also important that attention be given to moving from theoretical orientations, through frameworks and models, to explanatory theories. It is important that the assumptions underlying explanatory theories be articulated—assumptions about the nature of the human organism (stimulus bound, cognitive interactionalism, environmental determinism, nativism, and others). It is also important that the theories be testable, and that they be tested.

Following a period of sustained development, articulation, and test of individual theories, it would be appropriate for the field to compare theories by the deduction and test of strong inferences (for example, testing the relative strengths of Barker's (1968) behavior-setting theory in comparison to Willems' (1977) behavioral ecology theory, or Wicker and Kirmeyer's (1976) manning theory). It is believed that the impact of such tests would greatly advance the theoretical state of environmental design research.

SOME CHARACTERISTICS OF THEORY IN ENVIRONMENTAL DESIGN RESEARCH

While the purpose of this document is not to argue on behalf of any particular theory, criteria for the development of theory can be sketched. Environmental design research theory has emerged from a variety of intellectual directions, but there is a convergence of opinion on what constitutes adequate theory in the field (see, for examples, Patterson, 1977; Proshansky & Altman, 1979; Rapoport, 1973).

Environmental design research subscribes to the general scientific position that explanatory theories should be explicit, internally coherent, parsimonious, account for existing data, and lead to falsifiable hypotheses, the testing of which will refute or corroborate the theory. In addition, there are nine other characteristics of theory that are unique to the orientation of environment-behavior research.

1. *Value-explicit.* As the field is problem-centered and action-oriented, the values and assumptions behind particular conceptualizations and theories should be articulated.

2. *Human experience and action in the everyday physical environment.* A fundamental tenet of the field is the emphasis of research and theory to investigate events, actions, and people's experiences in their everyday settings.

3. *Quality of life.* Theory should address issues germane to the quality of life and the role of the physical environment in the quality of life (President's Commission for a National Agenda for the Eighties, 1980).

4. *Mechanisms linking organismic and environmental forces.* Since the beginnings of the field, behavior in real environments has been seen as a joint product of organismic forces and situational factors involving social rules of the environment. Organismic factors include physiological and psychological processes of individuals and social processes and cultural value systems of groups. Environment factors are multidimensional and holistic, including physical components, social components, and cultural components. Theory must account for the mechanisms linking these organismic and environment factors.

5. *Different scales of analysis.* Environmental design theory should address data at different scales of the physical environment, and should be imbedded in conceptualizations in the larger context of the socio-cultural, economic, and political environments. Environmental design research historically has focused on meso-scale environments—buildings and open spaces. This is an artifact of the origins of the field and its members, who have come largely from psychology and architecture. Conceptually, however, environmental design research pertains to all levels of the environment, from the macro-scale of large geographic regions, cities, and neighborhoods, to the micro-scale of objects in the everyday physical environment, such as street signs, and building materials (Saarinen, 1976). General theories need to respond to all three scales: macro, meso, and micro. At the macro-scale, theory needs to account for group behavior within large-scale environments (urban areas, landscapes, communities). At the meso-scale, theory needs to account for individual and group behavior within intermediate-scale environments (buildings, urban spaces). At the micro-scale, theory need to account for individual behavior and small-scale environmental units (ambient light, noise, stimulus complexity). The interests of portions of the field are focused on different scales (for example, social geography on the macro-scale, architectural research on the meso-sale, human factors on the micro-scale); these three scales, however, must receive equal emphasis in considerations of the overall relationship between behavior and environment.

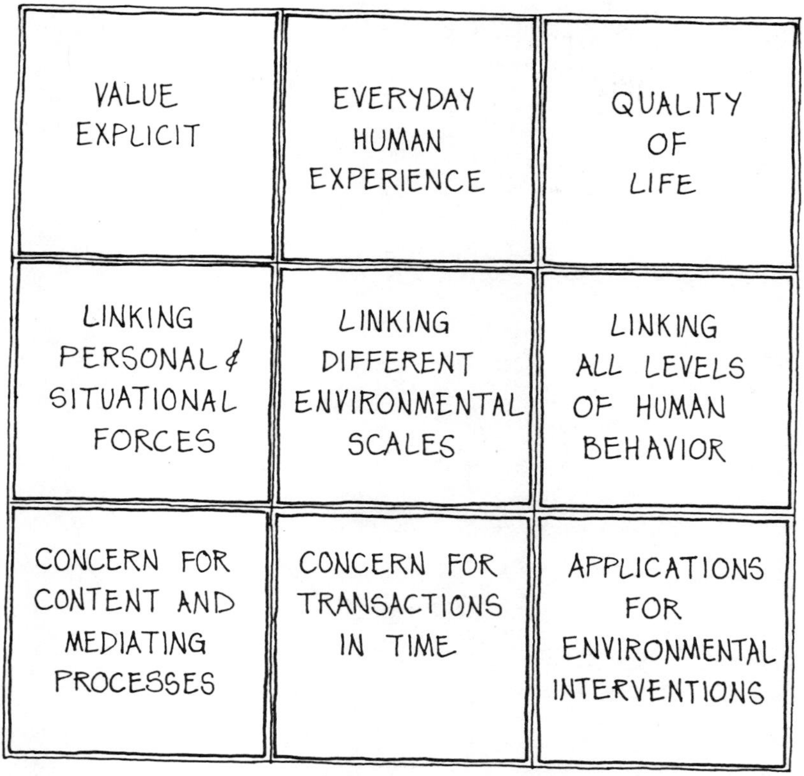

VALUE EXPLICIT	EVERYDAY HUMAN EXPERIENCE	QUALITY OF LIFE
LINKING PERSONAL & SITUATIONAL FORCES	LINKING DIFFERENT ENVIRONMENTAL SCALES	LINKING ALL LEVELS OF HUMAN BEHAVIOR
CONCERN FOR CONTENT AND MEDIATING PROCESSES	CONCERN FOR TRANSACTIONS IN TIME	APPLICATIONS FOR ENVIRONMENTAL INTERVENTIONS

FIGURE 16 Unique characteristics of good theory in environmental design research.

6. *Different levels of behavior.* The multidisciplinary character of environmental design research implies that theory should ultimately be able to deal with all levels of "behavior" in the broadest sense of the word. Stokols (1977a) proposed a preliminary profile of the environment-and-behavior field as composed of "intrapersonal process" and "environmental dimensions." He then suggested that within the total field of environment-behavior research, studies of individual behavior are most often conducted in "environmental psychology"; studies of small group behavior in "ecological psychology"; and studies of community phenomena by that portion of the field identified with environmental sociology, social geography, and urban anthropology. However, the multidisciplinary field of environmental design research must develop theory that accounts for data at all three of these different levels of human behavior. Theory should be able to account for physiological responses (for example, fatigue, arousal),

psychological responses (perception, cognition), individual behavioral responses (territorial behavior, privacy regulation), social group responses (small group dynamics, proxemics), and socio-cultural responses (neighboring, cultural values), to name just a few.

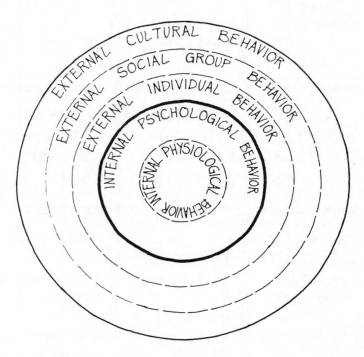

Figure 17   Levels of behavior in environmental design research.

7. *Content as well as mediating processes.* It has been mentioned in Chapter 3 that environmental design research is as concerned with the content of environment-behavior transactions (for example, with actual images of specific environments) as it is with the processes of the interaction and the mediating processes that intervene between organism and environment (e.g., the process of perception of the environment). Environmental design theory, thus, should be able to account for both specific content and mediating processes.

8. *Time.* Time is critical to understanding the ongoing transactions of people and their environment, and understanding the differences between short-term and long-term effects of environment-behavior relationships. For one example, among many, the ways in which individuals regulate privacy in relation to social interaction varies greatly over time.

9. *Possiblities of application to environmental interventions.* As mentioned above, two reasons for the importance of theory in environmental

design research are to aid in the applications of research findings in the environmental design professions and to aid in education. Both functions can be made more effective by presenting coordinated bodies of information and major concepts. Subfields of environmental psychology, environmental sociology, and other fields more closely associated with the social and behavorial sciences, tend to be oriented to basic research without explicit attention to applications. The field of environmental design research, however, is problem-centered *and* theory-centered. An important function of environmental design research theory is to suggest ways in which the physical environment can be changed (or prevented from changing) for the enhancement of the quality of life (Willems, 1977).

OTHER CONCEPTUAL ISSUES FACING THE DEVELOPMENT OF ENVIRONMENTAL DESIGN THEORY

In addition to the development of explanatory theories, there are other conceptual issues that are unique to this field which need exploration. Eight conceptual issues stand out as being of major importance for the continued conceptual development of the field.

Emergent Conceptual Ideas

In one of the early papers in the field, Proshansky (1972) called for the development of emergent conceptual ideas and research methods, that is, for the development of theoretical constructs uniquely suited for describing and explaining environment-behavior phenomena and applications.

Recent reviews of portions of the literature (for example, Stokols, 1978, 1981a; Russell & Ward, 1982) have identified a number of new concepts that have been developed that pertain specifically to environment-behavior interactions and their applications to environmental intervention (for example, fundamental vs. macro-spatial cognition, under- versus over-staffing, perceived environmental quality index).

Stokols (1978) suggested that the field has moved beyond the applications of established social science theories to the formation of new concepts and models pertaining to environment and behavior phenomena. For continued conceptual development, the field needs to continue to look for more powerful and useful conceptualizations from intellectual parent disciplines in both the social sciences and the environmental professions (for example, the theory of social behavior of Fishbein & Ajzen, 1975; cf. Weidemann, Anderson, Butterfield & O'Donnell, 1982). And yet there is a question as to how far one can take social science theories that are intended to account for person variables and interpersonal processes. Fundamentally, the field needs to continue to study and understand the integrity of environment-behavior transactions in their own terms (for example, Seamon, 1980b).

Resolution of the Theoretically Focused Views and the Problem-Centered View

The theme of many conferences has been some aspect of "bridging the gap" between research and research utilization. Smith (1977) has discussed the issue in conceptual terms. On the one hand, various psychologists and social and behavioral scientists have attempted to establish a framework that would give proper attention to the environmental contexts of behavior. While developing elaborate conceptual and methodological schemes, these traditions have paid considerably more attention to personal factors influencing behavior than to the physical environmental context. An exception is Barker's (1968) work, which can be viewed as a reaction against the intraorganismic orientation (but then it may go too far in the direction of environmental determinism).

There are problem-centered approaches on the other hand that contribute through research and application to the development of a more livable physical environment (Zeisel, 1981). Various research architects and environmental professionals have attempted to devise systems that would utilize environment-behavior data in the service of environmental interventions (for example, Alexander, 1964 and passim). While developing operational systems for utilizing research information, these approaches have not contributed to the conceptual development of the field. The challenge now is to conduct research that contributes equally to the development and articulation of theory and to applications to solving environmental problems.

Assessment of Different Research Methods

The field has had a commitment to studies of behavior in situ (Proshansky, 1972). Greater reliance on qualitative field research methods, including systematic observation and focused interviews, than on quantitative, laboratory research methods, seems in order. Yet the scientific training of most members of the field has been in the social and behavioral sciences (the largest number in psychology), where quantitative, experimental methods are stressed.

The result has been criticism on both sides. Those with rigorous training criticize studies from the qualitative side as being open to massive problems of internal and external validity. Those with applied interests criticize studies from the quantitative side as addressing variables that are inconsequential to everyday experience. A resolution adopting several of the quasi-experimental research designs of Campbell and his colleagues has been gaining recent acceptance (Campbell & Stanley, 1963; Cook & Campbell, 1979; cf. Patterson, 1977). Meanwhile, some scholars believe that the field needs more research from scholarly, historical, and ethnographic directions (for example, Rapoport, 1977; Stea, 1978). The field would benefit by an in-depth analysis of the pros and cons of all these approaches to environmental design research. It is unlikely that any one approach is "better" than any other. The field waits for a broad-based analysis of the appropriateness of different research methods (research designs and information gathering methods) for different contexts and types of questions.

Refinement of Definitions of Environmental Quality and the Quality of Life

Environmental quality and its role in the quality of life is the heartland of environmental design research. With changes in political and economic structures, the focus has shifted from the well-being of consumers to the efficiency of the production process. It is important to keep in mind the many social and environmental criteria by which quality— both environmental and the quality of life—is perceived in the public eye. Continued attention needs to be given to the issue of defining the quality of life and environmental quality and the relations between them. What are to be the critical variables by which the quality of the environment is to be judged? What aspects of human functioning are most related to the quality of life? How is the quality of the environment related to the quality of life? Some recent research looked at the possibility of developing "perceived environmental quality indices" (Craik & Zube, 1976). Considerable research could be focused on this direction, including further development of perceived environmental quality indices and development of user appraisal systems.

Measuring the Physical Environment

Much of the social and behavioral sciences categorize "environments" in either behavioral or perceptual terms. One of the tasks facing environmental design research is the development of a taxonomy of environments and of environmental variables that highlights the physical dimension. Some beginnings have been made recently to develop these taxonomies (see, for examples, Lazarus & Launier, 1978; Wohlwill & Kohn, 1976; Mischel, 1977; Moos, 1973, 1976), but still these discuss the social environment more than the physical environment. A central task of the field remains to specify structural relationships between attributes of environments and relevant aspects of behavior. The task can most effectively be carried out—for purposes of a first approximation—by abstracting from the selected classifications of environmental settings and variables that can be shown to be of significance in the study of structural relationships and to the quality of life.

Three levels of analysis appear necessary. The first level would deal with spatial scales of the environment (Saarinen, 1976). The second level could be particular types of environments—a taxonomy of settings (in architecture this is known as a *building type analysis*). These taxonomies might be based on biological analogies and may need to be developed within a formal geometry allowing comparison and manipulation (see, for example, Mitchell, 1977). The third level of the needed taxonomy could focus on the definition of the characteristics of settings which have noticeable relationships with behavioral outcomes and the quality of life, that is, determining objective dimensions of the environment (Wohlwill & Kohn, 1976).

The classification of environments—by scales, taxonomies, and dimensions—is similar to the levels of environmental influence discussed by Bronfenbrenner (1977), and can have profound effects on the field. With a clear system of organization, research can better focus on the socio-physical environmental factors related to the quality of life. Factors relating to the person as well as motivational and value-related factors can be studied as intervening variables modulating environment-behavior transactions. Environmental concepts, like complexity and unity, that are difficult to specify in physical terms because they are subjective, can be more easily studied as intervening concepts between objective dimensions of the environment and behavioral outcome variables.

Research on systematic taxonomies and dimensions of the environment will also aid in environmental evaluation as the sampling of environments will be made easier and more systematic. Similarly, it will aid environmental problem solving and decision making as research findings will be clarified.

Limits of the Environment on Behavior

While the field has been studying particular relations between design and behavior since the early 1960s, more needs to be learned about the limits of human adaptation and the effects of prolonged stress. Similarly, more needs to be known about which environments (scales, types, and environmental dimensions) are most critical to behavior, and which groups are most impacted by these factors.

The Economic Dimension

It may be ironic that while ordinary people in ordinary environments are highly motivated by the economic environment and its subtle and not-so-subtle incentives and constraints, the field of environmental design research has failed to recognize the importance of the economic climate. Important conceptual questions include the tradeoffs people make between environmental quality and economics, the processes by which tradeoffs are made, the limits people put on economic expectations, and the relative role of economic factors versus design, planning, and management factors in environmental satisfaction.

The Cultural Dimension

The study of environments and the general conceptual underpinnings of the field can profit by cross-cultural research in a generic way—a comparative perspective on the full range of environment and behavior phenomena. The ethnic and cultural composition of much of the United States and other developed countries is changing rapidly. The projected increase in the Hispanic population in the United States, increasing pluralization, the increase in black occupancy of urban areas, and massive population shifts and relocations around the world all portend the need for the field to study environment-behavior phenomena in relation to culture.

THE IMPACT OF THEORETICAL AND CONCEPTUAL ISSUES

This chapter has emphasized the role of theory in environmental design research, and has suggested the need for immediate and sustained research devoted to the development, articulation, and testing of theoretical perspectives, frameworks, models, and, especially, formal explanatory theories of the middle range. The chapter has also called for attention to priority conceptual issues which can contribute to the underlying structure of the field. A decade of sustained investigations of these issues will provide a more integrative perspective for a general theory of environment-behavior relations and for applications to improving the quality of the environment.

NOTE

1. These distinctions were first presented in a series of conceptual panel discussions chaired by Irwin Altman at the Environmental Design Research Association 7th Annual Conference, Vancouver, March, 1976.

6

PLACE ISSUES:
SETTINGS FOR HUMAN EVENTS

Major research directions and future needs regarding the study
of places as settings for human events. The chapter is divided into
discussions of large-scale place issues (planning and public
policy decisions affecting natural environments, regional areas,
and metropolitan areas), intermediate scale place issues
(workplaces, housing, other building types, and places affected
by change), small-scale place issues (room and building subsystems), and conceptual issues of place (experience, history,
and the management of place).

A new major area of research in environmental design organizes the
field around places, settings, and environmental variations.
   The framework of environmental design research discussed in
Chapter 4 suggested that any aspect of the field may be looked at in
terms of place, people, and the socio-behavioral issues that link people to those places. In earlier years, contributions to the literature
tended to be organized in socio-behavioral categories, along the
lines of traditional social and behavioral science categories like
privacy, personal space, environmental cognition, and attitudes and
preferences (see the EDRA *Proceedings*, 1969 on). However, a review
of research at the end of the first decade of EDRA (Ross & Campbell,
1978) showed a shift in emphasis to organization of place types. The
reviewers also noted that more research had been conducted on the
building scale than on either larger urban settings or interior

This chapter owes much to the assistance of Ellen Bruce, who worked with the
authors in conceptualizing the material and wrote the first draft.

spaces, and that housing was the single most studied setting. Facilities like airports, parks, child care centers, nursing homes, and libraries had been studied infrequently. The trend continues; respondents to the 1980 EDRA Survey (see Appendix B) suggested that research most needing attention should be organized around different types of physical settings.

This trend seems to address the need for a practical and integrative approach. For the practicing professional in environmental design, evaluation, and management whose work is often organized in terms of place types (interiors, buildings, or urban settings), research that is organized by place can be convenient for finding relevant behavioral information quickly, for analyzing critical environments, and for generalizing this information in professional practice. Organizing issues by place provides an orientation that design professionals can share with researchers, thus increasing the potential for communication and understanding.

In this monograph, the discussion of place research is organized into three categories of scale from larger-scale issues affecting regions, through the intermediate scale of buildings, to small-scale issues of interiors, products, and materials (Saarinen, 1976).

The divisions between scales may seem arbitrary, but they are the same general categories used in the environmental decision making professions. Large-scale spaces are typically the province of resource managers, urban planners, landscape architects, urban

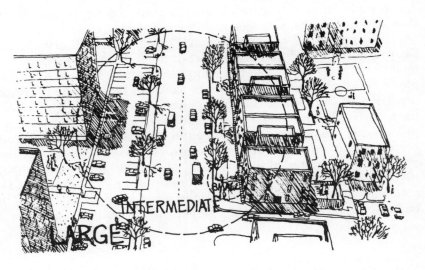

Figure 18 Nested hierarchy of scales.

designers, and civil engineers. The intermediate scale is typically the province of architects, building designers, and consulting engineers. And the smallest-scale issues fall within the province of interior designers, product designers, industrial designers, and graphic designers dealing with building subsystems and materials.

Although distinct in terms of applications, these scales of the environment may also be described as nested, one within the other. Rather than separate places with firm boundaries, a nested hierarchy of places allows for research across scales. For example, the study of user satisfaction in housing can be more significant within the context of the building's location in a neighborhood and city and its proximity to facilities and services. Satisfaction is better understood when also related to individual room sizes, layout and doors, windows, furniture, and other physical attributes. Overall satisfaction (or other measures of behavioral response) may be seen as the result of the relative contributions of issues identified at the large, intermediate, and small scales. When viewing the issues holistically, the relation between the various scales of places can take on a more practical and conceptual significance.

LARGE-SCALE PLACE ISSUES

For professionals concerned with large-scale environments, empirical information about the cultural context of these environments and the cultural and social consequences of large-scale planning decisions can aid policy and decision making. In addition to providing empirically based research for logical decision making, research at this scale can provide designers, policy planners, and decision makers with a working knowledge or framework for understanding critical relationships between people and space.

Three large-scale place research issues have emerged from the survey responses as important for the future (see Appendix B). The first is research on the relationships of people with natural settings, which has implications for regional planning, landscape architecture, and resource management. The second focuses on regional and rural issues, the relationships of current physical and social changes, and their possible effects on the quality of life within those settings. The third area of study deals with metropolitan and urban places and the physical, social, and economic trends that affect ways people experience and use their environment.

Figure 19 Natural landscapes (based on a drawing by Esherick, Homsey, Dodge and Davis, Architects, from Kemper, *Presentation Drawings by American Architects*, 1977)

Natural Environments

Natural landscapes have attracted considerable interest in the past 20 years. The growing public awareness of environmental issues is evident many ways: the significant increase in outdoor camping and recreational activities; the popularity of ecology and conservation movements; books and journals published on the subject of outdoor life; and rising memberships in clubs and organizations that center on environmental concerns.

These concerns for natural landscape issues have not been limited to public awareness. A growing number of nations are passing environmental legislation for the protection and management of the natural landscape as a valuable resource. These concerns are no longer the interest of only Western society but have now been recognized by developing nations as well. Worldwide attention has been focused on issues of environmental protection, preservation, and conservation. World conferences on the environment, United Nations committees, and environmental conservation organizations all address these issues (Zube, 1980c). However, attitudes and values have often identified the natural landscape as another resource for human exploitation, whether for exploitation of natural resources for production, or preservation of outdoor scenic landscapes for recreational use. Consequently, most of the natural environment is in the process of being planned, designed, and managed for human use.

Significant environmental design research has been done on all of these issues. For example, a major volume in the area of environmental assessment, *Landscape Assessment*, by Zube, Brush, and Fabos (1975), presented research on values associated with landscape settings, human responses to the visual landscape, and models and applications that have been developed for landscape planning and management. Other major compilations of essays in the area of natural environments are *The Interpretation of Ordinary Landscapes* (Meinig, 1979), *Landscapes: Selected Writings of J.B. Jackson* (edited by Zube, 1970), and *Our National Landscape* (Elsner & Smardon, 1979).

Natural landscape and resource issues often include social questions (Zube, Brush & Fabos, 1975). Rationalization of landscape protection based solely upon the concepts of environmental determinism, conservation of natural resources, or needs for outdoor recreation have not been successful in protecting the qualitative values in landscapes. Assumptions about how people use and adapt to the natural environment are an important part of natural resource management. Research on the relationship between people and the natural environment is sorely needed to clarify the social nature of resource management and to examine other significant areas of concern. We need to know more about aesthetic assessment of landscapes (including uniquely beautiful settings as well as the everyday landscapes in which most people live, work and travel), the use of natural landscapes and social attitudes, and assumptions about natural landscapes.

In particular, the aesthetic assessment of landscapes is of significant interest in policy and decision making (Craik & Zube, 1976). Planners, landscape architects, and other professionals involved in planning and developing coastal environments need a reliable means of identifying qualitative variables important in determining aesthetic quality consistent with the public's aesthetic values.

Another issue concerns the mental and physical health consequences of exposure to natural environments. Still other examples of needed research include: the human use of wilderness areas; how regulations affect use patterns and environmental quality; research on the meanings of different landscapes and what values people associate with them; and studies of the use and meaning of open space in the context of changing urban, neighborhood, and rural populations.

Finally, while development of theory at the building scale of environmental design research has begun (most of the theories mentioned in Chapter 5 apply to this scale), we also need to develop a theoretical understanding of the basic processes that relate people and cultural behavior at the regional and landscape scale.

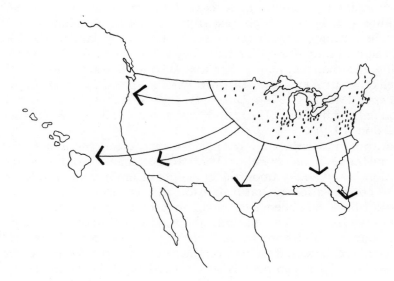

Figure 20   Regional migration shifts in the United States, 1970-1980.

Regional Areas

Recent population shifts and migration trends, from the United States northeast to the Sunbelt and away from large urban areas, have profound consequences for the quality of life in areas of population influx and those of population loss (U.S. Bureau of the Census, 1980).

The rapid development of the southeast and southwest in the United States and the phenomenon of boomtowns associated with natural resource exploration in places like northern Canada, the Rocky Mountain regions, and Alaska are creating profound physical and social impacts. Housing demand, overcrowded schools, and overstressed physical facilities are a result of the pressure of population growth. In areas of high growth, these processes can lead to conflicts between different subgroups in the population and to environmental degradation. An important research area, therefore, is the human consequence of rapid growth and alternative ways of handling change to ameliorate negative social, environmental, and economic costs (compare, Downs, 1970).

The urbanization of suburbia has given rise to successive population movements to adjacent nonmetropolitan territories. The growth of medium-size cities is a phenomenon that is interesting, not

Figure 21 Metropolitan areas (based on a drawing by Rutledge, *Anatomy of a Park*, 1971).

only for the effects of such growth, but also for the causes of it. Research into why medium-size cities have become attractive to more people can tell about emerging environmental preferences and attitudes that may become increasingly important as social-environmental phenomena.

Metropolitan Areas

Urban areas are human as well as physical systems. Changes within them constitute environment-behavior issues. The study of cities can focus on the interaction of the people with the physical environment in light of new social conditions, economic constraints, and technological changes.

Two of the most widely cited books on the topic of urban environments and human behavior are Jacobs, *The Death and Life of Great American Cities* (1961), and Michelson's *Man and His Urban Environment* (1976). Michelson's book discusses a number of social and behavioral issues at the urban scale, including life styles, changes over the life cycle, social class, socio-economic differences, values, and pathology. Much of what has emerged relative to social behavior in urban areas is limited in both scope and time (Michelson, 1976). More information is needed about combinations of social factors in influencing behavior with respect to the environment (for example, changes over the life cycle in the use of the environment by different social groups).

The decline of urban infrastructures, public transportation, roads, social services, and general maintenance is linked to national and regional population and employment shifts as well as to political and economic trends. Research on the effects of this decline on mobility, satisfaction, and general behavior in the urban environment is crucial if we are to understand the consequences associated with these broad changes. An important research area is the legibility of the urban environment in providing a sense of scale, image, and human identity to its users (Appleyard, 1969, 1981). Another concerns the form and management of extensive city environments for public value (Hack, 1984; cf. Bacon, 1967; Barnett, 1974).

INTERMEDIATE-SCALE ISSUES

More research has been done on buildings and intermediate-scale open space than any other scale (Ross & Campbell, 1978). Still, there are environments that will be particularly important in the coming years. They deserve research attention—workplaces, offices, environments impacted by rapid social and technological changes, commercial facilities, and other building types need study.

Workplaces

Concern for worker productivity, reindustrialization, and the growing percentage of office workers, along with a general interest in human satisfaction, has called for research that addresses the interrelationships between the human, economic, and physical aspects of the workplace.

A series of studies on offices was conducted by Brill and his associates (Brill, 1982; Sundstrom, Kastenbaum & Konar-Goldband, 1978; compare Gaskie, 1980). Starting from a critical assessment of earlier studies on office environments (Sundstrom et al., 1978), Brill conceptualized the relationship between worker satisfaction, worker productivity, and attributes of the office environment, and has shown a number of impacts that design can have on both satisfaction and productivity.

A second line of work is Becker's *Workspace* (1981), especially regarding social impacts of computerization (for example, transfer of work settings and activities to home environments). Another important study in the area of work environments is the evaluation of the federal office building in Ann Arbor, Michigan (Marans & Spreckelmeyer, 1982). Their recent publication, *Evaluating Built*

Figure 22 New workplaces.

Environments (1981), presents a systematic approach to designing and implementing post-occupancy evaluations (POE) through the presentation of an office building case study. While identifying a number of important findings about aesthetic assessments of the office environment, a major research need identified in this study is the need for better objective measures of job performance and worker satisfaction, and of the architectural variables that affect both.

Worker productivity is down in the United States. Despite the fact that this relatively short downturn is in the context of a long-run, upward trend, the problem looms as a major national concern. The relationship between the workplace and productivity must continue to be a central interest. While offices are now receiving considerable attention (Manning, 1965; Duffy, Cave & Worthington, 1976; Parsons, 1976), other workplaces—industrial plants, professional offices, electronics and communications firms—deserve equal time.

Worker satisfaction has become a critical social and political issue and is often seen as the intervening variable between the workplace environment and productivity. Empirical research on the relationship of these three factors could be useful to a wide range of professions and commercial enterprises that are eager for more information. Areas of research include the development of indicators that can measure these relationships and the development of appropriate research methods for particular workplace types (Wineman, 1982).

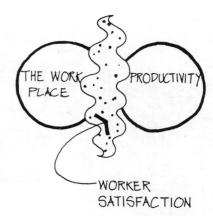

Figure 23 Place, productivity, and worker satisfaction.

As technological changes produce new types of information processing equipment and production systems for use in the workplace, consequences of these changes should be studied in order to do a human, as well as economic, cost-benefit analysis of technological change.

Housing

Although housing is more thoroughly studied than any other type of environment (see Ross & Campbell's [1979] review of the EDRA Proceedings from 1969-78), additional housing research is needed. Considering that housing is over 60 percent of all building constructed annually in the United States, housing research will continue to be a vital concern in the future.

Cooper's *Easter Hill Village* (1975) is a major example of a housing study in the field of environmental design research. More recently, the research conducted at the University of Illinois, Housing Research and Development Program (see, for examples, Weidemann, Anderson, Butterfield & O'Donnell, 1982) is an example of other leading investigations in the field. A number of issues raised by these evaluations of government-assisted housing suggest future research directions. First, can the findings and recommendations be generalized to other similar settings? Second, to what extent can findings of one case study today be applicable to future housing conditions? These questions are partly constrained by the complexity of the housing issue, partly by the methodology of housing studies, and partly by the limitations of resources available for research.

Figure 24   The adaptive reuse of old housing for new lifestyles (based on a drawing by Blessing, from Burden, *Architectural Delineation*, 1981).

There is also a need for continual housing research that develops and tests basic theoretical models of housing satisfaction from both longitudinal and cross-cultural perspectives. Research is needed that describes and develops a greater understanding of the basic relationships between environmental factors and users' satisfaction in different cultures and at different times. Some of the major volumes on these topics include Rapoport's *House Form and Culture* (1969) and Morris and Winter's *Housing, Family, and Society* (1978).

Social, economic, and technological changes occurring in the Western world, such as single-parent families, rising energy costs, and the competition for scarce land and oil, are new contexts for housing development and its research questions. For instance, research on the effects of new constraints on users' expectations and satisfaction with the environment could provide information leading to the development of alternative housing types.

As the housing crisis begins to affect the middle class, renewed concern about housing is being felt. Pending alternative housing types that could reconcile consumer expectations with present and future constraints will be important and have a considerable impact on the overall form of the environment. Housing needs to be evaluated in terms of the needs and expectations of ordinary people as well as special user groups such as first time owners and the elderly. The changes that must be made in existing housing to make it appropriate for use should be studied to find congruences with the needs and expectations of users.

Place Affected by Change

Many settings need examination in the light of rapid social and technological change and the impacts of changes on these environments and their inhabitants. Some examples of these places include reused existing urban buildings, airports, and technical research laboratories. It is important to study these types of environments because rapid technological change can affect their total character and the people within them. For example, in hospitals, the critical fit between people and equipment systems determines whether the environment and the staff function best for their own and for the patients' benefit (Carpman, 1983-84).

In another case, recent economic and social changes have made it necessary to adapt existing environments for different uses—what is known as *adaptive reuse*. Research on the reuse of educational facilities for new groups, for instance, can tell how school buildings can satisfy the new users' needs. Basic research is needed to identify which aspects of place are critical to achieve a successful behavioral fit with new uses.

Research on Other Building Types

Finally, there are a number of building types which have not been subject to much environmental design and behavior research. Notable among these are shopping centers, commercial facilities, transportation and communication centers, and information dissemination centers.

While almost 20 percent of the U.S. population changes residence each year, and this trend has persisted over two decades, changes in the economy resulting in decreased residential mobility (U.S. Census, 1980) have led to a focus on everyday environments, those settings which people use regularly such as gas stations, grocery stores, local shops, or neighborhood parks. Research on these places could have great impact if focused on simple behavioral fit and users' needs that could be incorporated into design guidelines. Examples of such design guidelines include *Planning for Cardiac Care* by Clipson & Werher (1974), *The Small Public Library* by Hill (1980), *Recommendations for Child Care Centers* by Moore, Cohen, and McGinty (1979), and *Low-Rise Housing for Older People* and *Mid-Rise Elevator Housing for Older People*, by Zeisel, Epp, and Demos (1978, 1984).

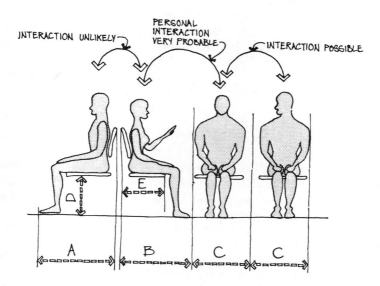

Figure 25 Human factors and interior design: the physical elements of daily encounter.

SMALL-SCALE PLACE ISSUES

Micro-scale issues deal with the environment-behavior relationships at the most immediate level of human interaction: those physical elements that we all touch, see, and use daily, like chairs, the work station, our offices, or kitchens. Ergonomics (or human factors) began with a concern for designing hardware to fit people, but research interests have broadened. A major—and early—study at this scale is Kira's *The Bathroom* (1966). Another example is Clipson and Werher's study, *Planning for Cardiac Care* (1974). These studies focused on the environment-behavior relationships of the smaller-scale settings, and identified those aspects of the physical environment having most immediate impact. Often these studies dwelt on the more practical aspects of anthropometric fit, and quantified physical reactions to sensory stimuli in the luminous, sonic, and thermal environments (compare texts by Bennett, 1977; Kleeman, 1981).

The studies of Hall (1966) and Sommer (1969) dealt with human responses to small-scale social settings such as table and seating arrangements, proximity, and personal space distances. However, more research is needed that relates the cultural and social aspects

of small-scale settings to specific environments, such as the work station in the office or factory, the kitchen, or school classrooms.

Human health, safety, and security emerge as important micro-
scale place issues with considerable research potential and need. Major research issues include toxic effects of building materials and
systems, accidents in different types of environments, planning and
design for safety and security (see Levin & Duhl, 1984).

CONCEPTUAL ISSUES OF PLACE

There are a number of questions that apply to places of any scale and
help to describe the nature and experience of place. The term place is
a rich one, as it has geographical and social connotations (Canter,
1977). The concept of place is more than objective dimensions that
define a space, walls of a room, and the physical objects that are
enclosed within it. Place can be understood as a psychological and
sociological phenomenon. Relph (1976) states that place is not just a
formal concept awaiting precise definition, but also an expression of
geographical experience. A few researchers, mostly emanating from
social geography, are beginning to study place as a phenomenological
experience—the direct experience of the world (Lowenthal, 1961;
Tuan, 1974; Rowles, 1978; Seamon, 1980b; Buttimer & Seamon, 1980).
Places are seen as "a complex integration of nature and culture that
have developed and are developing in particular location, and which
are linked by flows of people and goods (circulation) to other places. A
place is not just the 'where' of something; it is the location plus
everything that occupies that location seen as an integrated and meaningful phenomenon" (Relph 1976, pp. 3-4).

Experience of Place

Basic research questions cross-cutting studies of places at any scale,
large, intermediate, or small are: What makes space into place? What
are the qualities of experience that define place? What characteristics
of the environment, of human experience, and of cognition comprise
the essence of place? A task for environmental design research contributing to this discussion is to study place from the perspective of
the various users, how places are organized and used, and the
character of particular types of archetypical places: the home, the
workplace, the place of public assembly, the place of entertainment
and relaxation, the place of learning, places for being born, or the
place for dying. Some of the leading edge research into the
phenomenology of place comes from the area of phenomenological
geographers like Tuan (1974), Relph (1976), and Seamon (1980b).

The History of Place

Places offer insights into the historical character of a society. Research into the origin and evaluation of places can be difficult, handicapped by the comparative newness of the topic. Two researchers who have written eloquently about the nature and history of place and the significance of the evolution of place include Jackson, *American Space* (1972), and Clay, *Close-Up: How to Read the American City* (1973). Studied historically, the beginning of the occupation of land, the development of legal divisions and boundaries, the dwelling, the farm, the village or the town, all can reveal significant insights for future development, design, and planning. The interrelationships between behavior and the environment become more complex as the social, political, and economic growth of a place is investigated. The significance in such an area of investigation is not simply backtracking to describe history or particular places, but rather a historical perspective and theory on the nature and development of place.

The Management of Place

The management of place can affect its use significantly. As the quality of place is somewhat fragile, improper management may even destroy place. Conversely, the sensitive management of the environment can lead to the possibilities of its being used and interpreted as place. The relations between management and place are even less understood than the history of place. Some of the major works that begin to develop an understanding of the management of place are the writings of Lynch, *What Time Is This Place?* (1973), *Managing the Sense of a Region* (1976), and more recently, *A Theory of Good City Form* (1981). The field of environment-behavior research can contribute significantly to the development of theoretical models for environmental intervention and management to enhance the sense of placeness, the experience of place at all scales—regional, urban, neighborhood—and the smallest scale of places, such as our homes and workplaces.

PLACE AS AN ORGANIZING DEVICE

This chapter has emphasized the study of environmental settings as a critical organizing device for future environmental design research. Environmental settings have been seen in terms of scale (macro, meso, and micro) and in terms of conceptual issues of place cross-cutting these scales (including the experience of place, the history of place, and the management of place). A number of setting types have been identified that deserve particular attention, especially in the context of changing national priorities and technological developments (natural environments, regions, urban areas, workplaces, and alternative housing, among others).

7

USER GROUP ISSUES: PEOPLE IN PLACES

Major research directions and future needs regarding the study of different user groups. While recognizing the "big three" of children, the elderly, and the handicapped, a new conceptualization is offered that could organize research around groups affected by altered life styles, special life cycle requirements, highly impacted by the environment, and underrepresentation.

New orientations in environmental design research are required to ensure that future environments will be supportive of the style and quality of life of the people who use them. The increasing diversity and mix of people in society requires environmental adjustments because of constantly changing lifestyles. People change and therefore need settings that are adaptable and versatile enough to accommodate change.

The traditional environment-behavior classification of people into "user groups" such as children, elderly, or handicapped may create an insular approach to each issue. A more useful concept recognizes the interrelationship of all these groups within a larger economic, political, and social framework. Such an approach acknowledges design topics relevant to each group while integrating their various environmental requirements into a new approach to research. For instance, the understanding of children and elderly in the environment can be enhanced by seeing both user groups as part of a developmental cycle or by recognizing that environments often must accommodate the needs of multiple user groups.

This chapter owes much to the assistance of Patricia Gill, who worked with the authors in conceptualizing the material and wrote the first draft.

Life span developmental theories from psychology and sociology (for example, Baltes, Reese & Nesselroade, 1977) may be excellent starting points for research on the environmental attributes of human development and for the creation of a comprehensive theory of changes in relation to the physical environment. Examples of questions such an approach raises include: In what ways do the growth and development in children and the dissolution of functions in older people proceed in a similar but reversed sequence? How are the environmental supports and required stimuli similar and different? Do these related changes imply environmental correlates relevant for design, planning, and environmental policy?

Environmental user groups are classified herein according to larger social issues: altered lifestyle, groups with special life-cycle requirements needing greater environmental support, under-represented groups, groups highly impacted by their environment, and the group known as "ordinary people."

GROUPS AFFECTED BY ALTERED LIFESTYLES

A large portion of what is considered to be the average population has experienced extensive changes in social and political-economic structures within the past 20 years. These changes have forced alterations of lifestyle, that is, changes in style of living that reflect changes in the attitudes and values of the individual. Divorce, the changing role of women, and inflation are some of the causes that have affected traditional institutions like the family and the work force. Since these changes in lifestyle are occurring, the field should give serious consideration to defining and implementing associated environmental design research on the relationships between lifestyle and the environment (see Hayden, 1981).

The changing role of women in society is having one of the greatest impacts on cultures across the world and continues to be a vital concern for the future. Current research in the area of changing lifestyles of women, men, the family, and family life have been conducted by Hayden (1976, 1981) and Saegert (1980) among others. Some of the latest anthologies outlining the major issues and reporting on the latest research and research needs in the areas of women's changing roles and the environment include *Women and the American City* (Stimpson, Dixler, Nelson & Yatrakis, 1980), *Women, Culture and Society* (Rosaldo & Lamphere, 1974) and *New Space for Women* (Wekerle, Peterson & Morley, 1980). Priority issues incorporating the needs of women include research on

Figure 26 New lifestyles and the environment.

changes in building design, particularly the home and workplace, and research on changes in urban patterns of development due to women's changing lifestyles.

The changing role of women in society has not been limited to incorporating new roles in the home and workplace, but has altered the role of men and provided alternatives to the structure of the family. The women's movement has spawned men's liberation, allowing men a new look at their role in society. Thus, research is also needed on the new roles of men and their relationships to the environment.

The family unit is another good example of an institution experiencing change. Most design concepts are still built around a prototypical nuclear family in single-family detached housing. It is clear that one research area useful in policy making would be the special environmental needs associated with emerging family types such as single parent families, families with two working parents, extended families, co-parents and children. In families where parents are divorced, the environmental effects on the child of co-parenting and single parenting should be addressed. Research on the effects of living with two different people in two separate settings as compared to the traditional nuclear family may indicate how much diversity a child can experience in the environment and still maintain stability.

On a larger scale, studies are also needed that concentrate on the multiple responsibilities of men and women in single-parent and two-career families. Are these responsibilities aided or exacerbated by current urban structures? Accessibility to child care, health care, shopping, and urban facilities has to be investigated when considering parents who work while taking care of a family. Basic research questions revolve around the relation of different patterns of urban/suburban structure and different lifestyles. For example, to what degree do central-city and suburban land patterns based upon traditional nuclear families support lifestyles now seen in single parent or two-career families?

Changing lifestyles have placed new demands of the environment in other ways. What are the implications of a shorter work week, early retirement, and the increasing emphasis on leisure, travel, and outdoor recreational activity? Other issues relevant to changing lifestyles include: What are other desired lifestyles and which environmental characteristics are needed to support them? How adaptable are physical environments to the changes in role choices society has begun to encounter?

GROUPS WITH SPECIAL LIFE CYCLE REQUIREMENTS

As people pass through stages in the life cycle, behavioral patterns, and dependence on the environment shift rather dramatically (Lawton, 1980). People do not remain in any one stage long. While these facts seem obvious, they are frequently forgotten when policy, planning, and design decisions are made. Recently research has been devoted to certain stages of the life cycle, for example, the special needs of children and the elderly. Other stages, like adolescence and middle-age years, receive too little attention. Research on each stage, and more importantly, on life-cycle development in relation to the environment deserves continued encouragement.

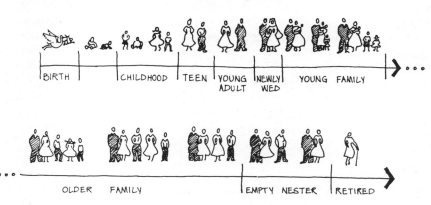

Figure 27 The relation of life cycle requirements to the physical environment.

Children

Often the environment has not been designed for children: There is essentially a misfit between the adult environment and the requirements of the young and small. As a result, children have more serious accidents than other people (bicycle and playground accidents rank among the top ten causes of accidents for all age groups; U.S. Consumer Product Safety Commission, 1980).

Advances have been made on research on children in natural settings, in home environments, in school settings, and on behavioral issues like environmental cognition and the regulation of privacy (Weinstein, 1979). Under the aegis of the U.S. Army Corps of Engineers, some of this research has been translated into planning and design guidance (Moore, Cohen & McGinty, 1979).

While research in this area has been expanding rapidly (Weinstein & David, in press), questions have been raised about the relative reliability and validity of much of it (Weinstein, 1979; Moore 1982a). Questions revolve around the use of one-shot case study research designs, inadequate calibration of measurement instruments, problems of small samples, and problems of external validity.

In the light of the environment's presumed effect on development and learning, additional research is needed about the interaction of children and the environment using quasi-experimental field methods. Some needs include examining the impact of indoor and outdoor play areas, the impact of house and neighborhood design, and the impact of formal learning environments in relation to home, neighborhood, and workplace environments. Equally important is the impact of the environment on mentally and physically impaired children, the relation of the designed environment to the amelioration of handicaps ("therapeutic environments"), the implication of whether children or adults make decisions about the use of environments for young children, and ways to involve children in the design and planning process. Underlying this area is the rhetorical question raised by Doxiadis (1974): How would the environment be different if it were designed from the child's point of view?

Older People

The fastest growing sector of our population is people over 70 years of age (Lawton, 1980; White House Conference on Aging, 1981). These changes will soon impact on the developing world as well. Previous assumptions about the numbers of older people in the world and ways in which the environment meets their needs are no longer applicable. Due to improved standards of living, including better health care and changes in life-styles, people are living longer. Behavioral issues associated with the elderly are no longer a matter of supporting more elderly people for a longer period of time, but understanding the environmental conditions associated with prolonging productive and meaningful life experiences—what is called the environmental supports for successful aging.

Some important new research information about the elderly is available. Carp's work, *A Future for the Aged* (1966), outlined some basic research and thinking in the area of the elderly and environment and is an excellent example of quasi-experimental research design. Research on aging and environment has also begun to find applications. Most recently the award-winning work of Zeisel, Epp, and Demos (1978, 1984) has provided planning and design guides for low-rise and medium-rise housing for older people.

Some priority issues that still need attention include: the impact of alternative housing configurations on the elderly; the relationships among buildings, services, and human well-being; and actual physical barriers and constraints in the environment that can be eased through design. Design research can help deal with such concerns as developing ways to facilitate movement through better planning and design, and ways to provide needed cues for locomotion, mobility, direction, and orientation by the use of signs and legible design. The relationship between the older person's well-being and housing size and layout is of critical concern. Physical adaptations to the needs of the elderly in large- and small-scale settings should be a basis for research, design, and policy. Greater understanding concerning the implications of shared facilities, the transition from private to congregate housing situations and the implications on offices of an aging workforce is needed.

GROUPS THAT ARE HIGHLY IMPACTED BY THE ENVIRONMENT

Groups that are highly impacted by their environment are those that have the least amount of control relative to their day-to-day functioning in it. They include institutionalized people or those who are mentally or physically impaired. Handicapped people who are living in the mainstream of society may also be included in this category because they are subject to environmental and physical barriers in their daily lives.

Figure 28   Groups that are highly impacted by the environment.

Environments are generally not designed to support the needs of this particular group. This is especially true for people who are mentally or physically impaired. Bold beginnings have been made, however, in research and applications on populations highly impacted by the environment. For example, recently completed research of Steinfeld and his associates (1979) has contributed to the development of revised standards addressing barrier-free design model building codes, municipal codes, and some international codes and standards (see American National Standards Institute ANSI A117.1, 1979). Another example is the award-winning work of Farbstein, Wener, and their associates (1979, 1982) who have been conducting research and developing policy and design guidelines in the area of correctional environments.

Many design research issues concerning environmental limitations of the physically and mentally impaired, as well as those user groups who are institutionalized, still need to be dealt with. Prime examples are the ways in which the physical environment can support normalization, the impact of the physical environment on impairments like mental retardation and emotional disturbances, and the role of the environment in helping people to overcome or minimize their impairments, i.e., the therapeutic or learning environment

(Canter & Canter, 1979; Friedmann, Zimring & Zube, 1978; Moore, Cohen, Oertel & Van Ryzin, 1979).

Other areas of research that require greater attention in the future include: the impact of de-institutionalization on communities; the implications of living in more intimate, decentralized environments within the mainstream of society on the well-being of developmentally disabled people; the encouragement of mainstreaming therapeutic institutions; and the study of physical features that should be included in an institutional environment to enhance residents' opportunity for maximum development.

UNDERREPRESENTED GROUPS

Groups that are poorly represented politically and economically are often mistakenly assumed to be unable to make decisions about their own environments. Minorities such as black and Chicano migrant farm workers and ethnic populations like Eastern European and Oriental immigrants historically have had problems being represented. Insensitivity to their needs is a serious problem involving the imposition of inappropriate values and environmental assumptions.

Figure 29 Underrepresentation and the environment.

Research needs to be focused on defining where value differences lie between underrepresented groups and the mainstream of society, including designers. Research that examines the degree to which people of other backgrounds can adapt to the general environment and the degree to which their traditional values should be included is needed. Answers to some of these questions can be arrived at through improved communication between professionals and clients. This research that promises to address user group involvement in planning and design should be encouraged.

ORDINARY PEOPLE

Much of the United States, Canada, and other countries is made up of suburbs, single family housing, commercial strips, and shopping malls stretching across the landscape. Design professionals and researchers alike have ignored these environments as unworthy of study. The suburban landscape is often dramatized as the result of middle class, unsophisticated design awareness. Its design and study is often left to the discretion of developers, contractors, real estate salespeople, and finance companies.

The situation is changing. Some of the most influential and leading research in the area of ordinary people and their physical environment is the work of Gans, which includes his seminal works, *The Urban Villagers* (1959) and *The Levittowners* (1967). More recently Gans has expanded his research to popular culture in the book, *Popular Culture and High Culture: An Analysis and Evaluation of Taste* (1974). Other recent research in the area of ordinary people includes the volume by Schwartz, *The Changing Face of Suburbs* (1976) and Tuttle's *Surburban Fantasies* (1983).

The material success of Western culture has been due in part to the growth of the middle class. More needs to be known about this group as a whole. What changes in the environment can be expected to occur due to the changing needs of this large group of people?

Traditionally the suburban environment has represented certain social patterns (the nuclear family, single family homes) that may no longer be as realistic or attainable as they once were. Can it be assumed that these types of environments are responses to beliefs and preferences of those who use them? What types of diverse environments must be created as alternatives to suburban settings? Research on ordinary people could have significant implications for urban form and the design of environments for developing as well as technological societies for decades to come.

Figure 30   Ordinary people.

NEW USER GROUPS

This chapter has summarized research developments, unresolved issues, and research priorities regarding environmental user groups. While acknowledging the continued need for research on children, older adults, and the developmentally disabled and the environment, a conceptualization is offered that stresses changing life-styles, special life cycle requirements, environmental impact, political representation, and forgotten groups.

8

SOCIO-BEHAVIORAL PHENOMENA: PHYSIOLOGICAL TO CULTURAL

Major research directions and future needs regarding behavioral and socio-cultural responses to the environment. A conceptualization is offered that stresses the expanding circles of influence from the inner, organismic, or personal response level to the external, socio-cultural, and cross-cultural response level. Important issues regarding physiological responses to the physical environment, health and human welfare responses, psychological responses, individual and group responses, and socio-cultural responses are highlighted.

While the field of environment-behavior studies is experiencing a shift towards issues organized by types of place, settings, or types of environments (see Chapter 5), this chapter addresses a number of social and behavioral research areas that continue to need additional attention. One focus of empirical research has been behavioral and socio-cultural phenomena in relation to environments and application of this knowledge to environmental design. The Proceedings of 10 of the first 14 annual conferences of the Environmental Design Research Association were clustered around topics such as environmental perception, cognition, privacy, crowding, and personal space. These issues reflect interests of researchers trained primarily in the social and behavioral sciences with their highly conceptual systems for studying individual and group behavior.

Human behavior can be organized in expanding circles of influence from inner, personal responses to external, socio-cultural responses (see Chapter 5, Figure 17). Within these two categories further classification should be made. Internal responses include physiological and experiential-psychological responses. External socio-cultural responses include individual external responses and small group behavior, proxemics, neighboring, and responses of organizations and societies. Our review of the literature in environment-behavior studies has identified 20 prime areas of research which can be classified in one or another of these groups of behavioral responses. The system can provide a simple way to organize individual and socio-cultural behavior in order to identify major gaps in research.

INTERNAL PHYSIOLOGICAL AND PSYCHOLOGICAL PHENOMENA

Internal physiological and psychological phenomena refer to the processes inside the person including physical responses to external environmental stimuli like light, sound, heat, and cold, and mental responses, like vision and thinking.

Internal Physiological Responses

The major focus of study in the area of internal physiological responses has been the five human senses and kinesthetics (the sense of balance and movement). Also important are various states of physical health of the human organism. The relevance of such studies lies in identifying the impact of various environmental conditions on human health, comfort, stress, and other performance measures.

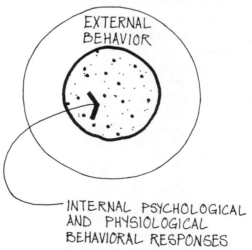

Figure 31 Internal physiological and psychological responses.

Considerable research has been conducted and a number of models developed that demonstrate physiological responses to the environment. For example, findings on noise indicate that noise exceeding certain levels can contribute to nervous tension, anxiety, and exacerbate effects of other illnesses, contributing to psychosomatic illness, reducing concentration, and affecting learning performance (Farr, 1972; Cohen, Glass & Singer, 1973; Glass & Singer, 1972). Sound, sound quality, and human response to sound has been thoroughly studied for applications in the recording industry, the design and construction of theaters, and for setting sound standards for work and living environments (for example, the California standard that requires every city in the state to develop an ordinance for noise abatement). There is also some research on the cultural variability in physiological response related to the design of the environment (Rapoport & Watson, 1967).

In relation to these internal physiological responses, there is a need to know more about health effects of various materials, equipment, furnishings, maintenance practices, and building configurations. There is a need also to understand which types of indoor pollution increase over time. With rapidly developing technologies of building materials, there is an increasing variety of potential health hazards that must be investigated (Levin & Duhl, 1984).

Internal Psychological Responses

Other internal responses include issues of perception, cognition, meaning and symbolism, psychological stress, privacy, development and learning, and emotions. There has been considerable research in environmental cognition (Moore, 1979b; Evans, 1980). The areas of study in meaning and symbolic interpretation have recently generated significant interest in the design community as well as research fields. The study of emotional responses to the environment is in its infancy and needs considerable attention.

Two major problems in the study of internal psychological responses are the confusion as to what these various terms mean, and the difficulty in studying internal, subjective states. Contributing to the confusion are the number of different disciplines using these terms, a number of different definitions, and arguments about methods.

As noted by several commentators, perception and cognition have been used in a confusing variety of ways (Downs & Stea, 1973; Moore & Golledge, 1976). Although the distinctions between perception and cognition fall short of forming a clear dichotomy, but rather fade into differences of degree and focus, certain distinctions can be made. *Perception* is the study of the sensory apprehension of objects and environments. It is generally defined by immediacy (that the behavioral response immediately follows sensation) and stimulus dependency (that a great proportion of the response can be accounted for by the physical properties of the stimulus (Moore & Golledge, 1976). *Cognition*, then, is the process of thinking and knowing the environment based on perception and other organismic and environmental factors. It may be conceptualized as a construct to account for the various means of processing information that intervene between sensation and subsequent behavior. Thus it includes the processes by which visual, linguistic, semantic, and behavioral information is encoded, stored, decoded, and used (Moore & Golledge, 1976).

Considerable work has been done on environmental cognition (see, for example, the anthologies by Downs & Stea, 1973; Moore & Golledge, 1976; cf. Moore, 1979; Evans, 1980). Most of this work has been at the urban scale; the field needs more understanding of how people form and use images of intermediate-scale environments, like orientation and wayfinding in complex buildings (hospitals, shopping centers, department stores).

Visual perception of the environment has also been of some interest, but to a limited number of scholars (for example, Appleyard, Lynch & Myer, 1964; Garling, 1970; Hayward & Franklin, 1974). The field has barely scratched the surface of the visual perception of the micro, meso, and macro environments. Architects, interior designers, and urban designers deal with the perception of three-dimensional forms every day in their professional work. The need to understand the processes of environmental perception, and the content of what is perceived under what situations, is paramount in the field.

Recent efforts both in environmental design research and in architectural theorizing have tried to bring the concepts of meaning and symbolism back into the arena of discourse (see Bonta, 1979; Broadbent, Brent & Jencks, 1980; Jencks, 1981; Rapoport, 1982). While there is considerable theorizing, more empirical studies are needed of repeated experiences across cultures at all levels of experience—perception, cognition, meaning and symbolism, and affect and imagery—and at all scales (see Groat, 1981; Groat & Canter, 1979). While other fields study one or another portion of this matrix of levels of experience by scales of environment, environmental design research can contribute to the discovery of an empirical basis for understanding the perception, cognition, and meaning of environmental experience focusing on the complexity of all of these subjective variables, thus providing testable theories of the designed environment.

EXTERNAL BEHAVIORAL, SOCIAL, AND CULTURAL PHENOMENA

While the "internal" phenomena discussed above concern the inner experiences of people relative to the physical environment, what are here called external phenomena concern the manifestations of these experiences. They occur at three levels: the behavior of the individual (for example, spatial movement patterns); the social interaction of small and large groups of people (such as neighboring); and the culturally based responses of ethnic groups and societal populations.

Individual Behavioral Responses to the Environment

This category deals with visible behaviors that are measurable. Within this category are behaviors such as spatial movement, orientation and way-finding, productivity, anthropometrics and ergonomics, personalization of space, personal space and territoriality, preferences, choices, environmental attitudes, evaluations and assessments, privacy, and crowding.

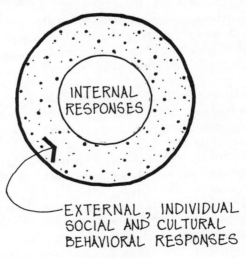

Figure 32 External behavioral responses.

Much environment-behavior research has been conducted on some of these topics and almost none on others. Sommer (1969) and Hall (1959, 1966) have highlighted the importance of personal space and interpersonal communication at different distances. Research on privacy, crowding, and density have been studied extensively by, among others, Altman (1975) and Stokols (1972a, b). Newman's (1973, 1980) work in the area of defensible spaces has gained recognition, not only from the perspective of environment-behavior research but architecture, planning, and urban design. Anthropometrics, the study of the dimensions of the human body, has been well documented by the work of Dreyfuss (1966) and used extensively in the design of small-scale environments and human factors engineering (the design of human-machine interfaces).

Environmental epidemiologists, health facilities planners, and human factors engineers strongly advocate that attention be given to new areas of research in performance and productivity. Environmental conditions related to performance and productivity in a variety of settings needs sustained research. Examples include the influence of temperature and light levels on comfort and performance, the detrimental effects of glare, the effects of visual and acoustic privacy, user control of environmental conditions, and the effect of selection and arrangement of furnishings and decorations on performance. Many conditions have been changed in work environments (for example, reductions and eliminations of the ability of workers to control or personalize their environments) without full knowledge of the impacts of these changes. Examples of this research would be analyses of worker health, attendance, and production records which could be compared with environmental qualities that fall within the environmental designer's scope of decision making. Assumptions of building owners and employers could be compared with actual performance. Case studies, interviews, analysis of work load and production changes could lead to a characterization of the critical environmental variables and subsequently to a significantly improved, data-rich basis for the design of work environments.

Some other topics of increasing importance are the relationship between environments and health, safety, and security. More is known about protecting buildings from fire, flood, and earthquakes than about protecting the occupants and getting them out safely. More should be known about which factors in building design and operation contribute to accidental death and injury. More should be known about the degree of risk associated with the use of improperly designed building elements. As concerns for the cost of energy increase, basic designs of buildings are changing without adequate understanding of the impacts on security, safety, and health.

External Social, Interpersonal, and Group Behavior

Research has focused on the environment and social group behavior such as group dynamics, proxemics, neighboring, and organizational behavior in relationship to the physical environment. Some of the key research in this area includes: the social life of small urban places (Whyte, 1980); the sociology of neighborhoods (Jacobs, 1961; Michelson, 1976; Saarinen, 1976); and social aspects of urban life (e.g., Altman, 1975; Mercer, 1975; Milgram, 1970; Duhl, 1963; Gans, 1962, 1967). There are many other studies that deal with the organization of behavior in groups and the relationship to urban form and the built environment—social patterns and the effect of neighborhood; relocation and population movements and their relationship to the environment; urban, suburban, and small town environments and their relationship to different social behavior of different groups. Priorities for research in this area include: consideration of not just the effects of environmental practices but also their own social causes and conditions; the sociology of the design profession and its effects on the built environment; the examination of major social factors of family pattern, life-cycle stage, and social economic class in relation to special features in the environment; the role of different lifestyles in affecting the character of the designed environment; and the "opportunity fields" for different social group behaviors provided by intermediate spatial scales.

External Socio-Cultural Behavior

The third level of external human behavioral responses deals with those aspects of culture that form the larger backdrop of human interaction with the environment. This includes latent and manifest cultural norms (Merton, 1957), rules and attitudes that affect human behavior individually or in groups, and the concepts of family structure, status, social and organizational hierarchy of roles, class structure, group identity, and cultural group behavior.

Socio-cultural research dealing with world views, values, life-styles, environment as communication, meaning, and symbolism is missing (some exceptions are Stea, 1978; Rapoport, 1976, 1979, 1982). These absent models would deal with people as members of groups and the nature of these groups, the criteria whereby people see themselves as a group and are seen by others as such, and distinct ones brought about through clothing, landscapes, or building. These models deal with how world views and values might affect specific ways of doing things and the mental images with which people communicate. More importantly, there is another question in all these matters: Which aspects are constant, which vary, and which are due to cultural differences? In all this, more cross-cultural and historical analyses are clearly needed.

HUMAN SOCIO-BEHAVIORAL ISSUES IN RETROSPECT

With respect to internal and external behavioral responses, it is interesting to note that we have been able to identify about 20 major behavioral response categories. Viewed within the frame of Maslow's hierarchy of needs (Maslow, 1954), many of them deal with the lower order of physiological needs: hunger, thirst, sex, activity, rest, and safety. The feeling of belonging and love, friendship, interpersonal relationships, and sense of group identification are also a part of this hierarchy. Less emphasis has been given to still higher-order needs on Maslow's scale such as self-esteem and achievement, competence and independence. While the highest order is the need for self-actualization, this need has been virtually ignored in environment–behavior–design research.

Figure 33 Maslow's hierarchy of needs.

9

PROCESS ISSUES: RESEARCH ON THE UTILIZATION OF KNOWLEDGE

The cyclical model of the processes of environmental policy planning, design, evaluation, and research is used as an organizing framework for this chapter. Research needs in each step are identified, including instrumental utilization, environmental programming, computer-aided design, participatory design methods, new approaches to post-occupancy evaluation, and quasi- experimentation and multi-dimensional analysis techniques.

Knowledge utilization does not occur in a vacuum. The utilization of scientific information in the formulation of public policy, even under ideal conditions, is the result of a complex and often seemingly capricious set of circumstances; some of these vicissitudes, however, are foreseeable.

Nathan Caplan
The Use of Social Science Knowledge in
Policy Decisions at the National Level, 1975

It is recognized in this book that environmental design research, while a scientific discipline, is also problem centered and applied. While pursuing interesting lines of inquiry that promise to lead to more fundamental knowledge about environment-behavior transactions, the field is also concerned with making the physical environment a better place in which to live.

A central concern in the field, therefore, is the processes by which knowledge may be used to increase the effective application and utilization of environmental design research in all related professions charged with the planning and design of our environment.

This chapter reviews priority research on the methods of environmental design research and on the ways it is applied and used in solving environmental problems. It is not the purpose of this chapter to discuss ways in which design research can actually have more impact on environmental decision making. Rather, it is to highlight some of the questions about these application processes that need additional research in order to better understand them. In fulfilling this more limited purpose, the chapter will highlight needed research on environmental policy planning, facility programming, computer-aided design, and other design methods. The overall planning and design process, as well as needed research on the process of environment-behavior research itself, will also be discussed.

The intent of this research is to improve environmental planning and decision-making methods and to understand how to use research information in shaping people-oriented environmental decisions. This can be achieved in three ways. First, research on process issues may increase the knowledge that actors involved in planning, design, and building have the criteria for their decisions. Second, process research may help improve the effectiveness of design and environmental planning institutions. Third, process research may affect these institutional and individual processes so that they may be more receptive to information generated by the field of environmental design research.

A CYCLICAL MODEL OF THE PROCESSES OF ENVIRONMENTAL PLANNING, DESIGN, EVALUATION, AND RESEARCH

As discussed in both Chapters 3 and 4, environmental design research is an iterative process that borrows knowledge from a variety of social and behavioral science fields and contributes to a variety of environmental planning and design fields.

The use of research in environmental design has been conceptualized as a cycle (Conway, 1973; Zeisel, 1975; Moore, 1979c; Villecco & Brill, 1981). Those models illustrate occasions for the application of environmental design research in the design process.

Three alterations are offered to this general model. First, construction is not considered the domain of environmental design research. Research on this phase of the design-build cycle falls within the area of construction technology and the building sciences (Snyder, 1982). Second, environmental design research extends beyond evaluation and programming research. It includes basic and applied research, which sometimes gain motivation from questions raised in post-occupancy evaluation (Moore, 1979c). Third, and more

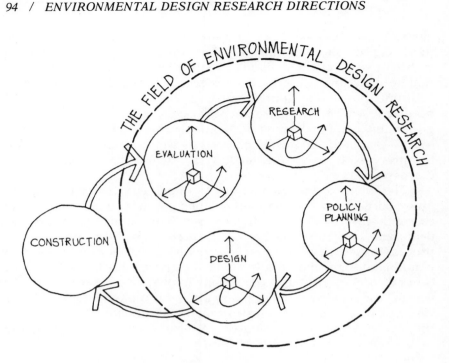

Figure 34   The planning and design cycle.

fundamentally, environment-behavior research is also used in the planning process (see, for examples, Lynch, 1960; Schorr, 1963; Lansing, Marans & Zehner, 1970).

The cycle of environmental design research and application consists of the following stages: (1) *environmental policy planning* (problem identification, policy formulation, goals and objectives, formulation and evaluation of program alternatives, and plan implementation; compare Zube, 1980a); (2) *environmental design* (programming, schematic design, design development, construction documents, and construction supervision; compare Palmer, 1981); (3) *evaluation* (defining the problem and research questions, designing the evaluation, conducting it, and analyzing the data; compare Friedman, et al., 1978; Bechtel, n.d.; Reizenstein & Zimring, 1980); and (4) *research* (including basic research and applied research, including designing the study, developing research methods, analyzing, summarizing, interpreting, and translating the results; compare Michelson, 1975; Zeisel, 1981).

Research on the utilization of design research knowledge thus includes systematic study of each of these steps as well as the study of implementation strategies between each step. The remainder of this chapter is devoted to exploring research needs in each of these domains.

ENVIRONMENTAL POLICY PLANNING

Environmental design research information is not limited in application to the meso-scale design professions (that is, architects, building designers, landscape architects, interior designers, product designers). Findings from environmental design research have direct application to decision makers in the policy-planning professions as well (that is, urban planners, environmental managers, politicians, and public policy analysts).

The term *planning* has different definitions and connotations. Part of the difficulty in defining planning for nonplanners lies in the confusion between policy planning and physical planning. Where does one begin and the other end? There is also economic planning, transportation planning, social planning, development planning, and neighborhood planning (Beal, 1968; Alexander, 1981).

Alexander (1981) defines planning as a process of "deliberate social or organizational activity of developing an optimal strategy of future actions to achieve a desired set of goals for solving novel problems in complex contexts, and attended by the power and intention to commit resources and to act as necessary to implement the chosen strategy" (p. 7). Even this is a broad definition and may encompass aspects of policy planning as well as design activity.

In many respects, the process of environmental policy making is similar to the design process. The design subcycle of programming, design, construction, and building evaluation is paralleled within policy development by problem identification, policy formulation, implementation of action programs, and program evaluation. In many cases environmental policy development in planning precedes facility programming in the design cycle. These plans provide general goals in which design decisions must be made.

Often the difference between design and policy development is a matter of scale. Policy often deals with larger-scale general issues, while design deals with more specific and concrete issues. For example, policy development and planning can establish goals and objectives for urban design while design carries out these general guidelines in schematic form and construction documents (Barnett, 1974).

In another respect, policy making is inherently different from design. Policy deals in a general and abstract way with establishing

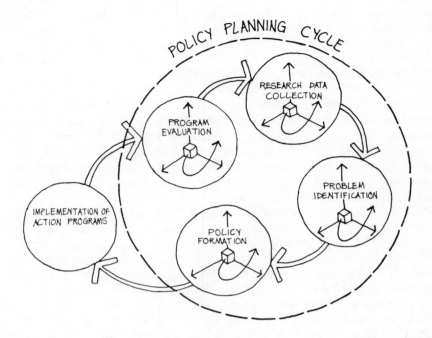

Figure 35. Environmental policy planning.

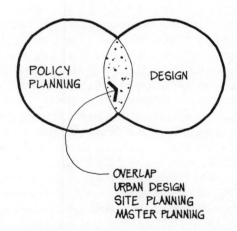

Figure 36. Policy and design.

goals and objectives. Policies are stratements that encourage, coordinate, and provide guidance for environmental issues. Environmental policy, therefore, is a statement of general intentions of a decision maker or a decision-making organization about the physical environment. These policies provide a context in which more concrete design decisions of the landscape architect, architectural designer, or site planner are made.

Environmental design research information may be used in all activities of policy making, physical planning, and design. A key example at the national level is the environmental policies embodied in the U.S. Department of Housing and Urban Development's Minimum Property Standards. These standards include statements about site planning that are actually statements about ways that planning and design can contribute to the quality of life. Recent research has addressed some of these issues, and revisions to the standards have been recommended (see Cooper, 1972).

Other aspects of policy planning fall outside the sphere of interest of environmental design research. For example, economic policy development or transportation policy development that do not deal simultaneously with physical environmental and human behavioral issues are not considered environmental design research.

The value of research in policy development is in improving the decision-making process, enabling it to be based more on information than on conjecture. Research on processes of policy making can provide researchers with an understanding of decision making, allowing them to better advocate environment-behavior concerns.

Because of the applied and action orientations of the field of environmental design research, the role of knowledge and use of scientific information in decision making are very important. The role of knowledge in policy formation has been studied in relation to public analysis, planning, and administration (Wilensky, 1976; Caplan, 1977; Weiss & Barton, 1979).

Of several models evaluated, the predominant model for the utilization of social science information at the national level is one of *instrumental utilization* (Caplan, Morrison & Stanbough, 1975; Weiss, 1977). Instrumental utilization is a pragmatic concept that suggests ideas are instruments applied in decision making as guides for action, their validity being determined by the success of that action. The purpose of instrumental utilization is to motivate decision-making interest in the use of research in environmental interventions. Characteristically, the utilization of research is indirect and bound to undergo further decision processes, distributed over numerous levels, and with delayed processing of the results. The final result is often a low visibility of research application.

There are many reasons why instrumental utilization has found research information not to be effective in national policy formulation: the frail character of the knowledge that social science research produces; the lack of clarification of policy choices; the implicit values embodied in social science research not being congruent with decision makers; the nature of the decision-making process and researchers' lack of knowledge about how decisions are made; and the mistaken assumption that identifiable actors exist who "make the decisions" (Weiss, 1977).

A number of characteristics have been identified, however, for the successful utilization of research (Caplan et al., 1975): decision makers should be oriented to and appreciate the scientific aspects of policy issues; ethical and scientific values of the policy makers should be oriented toward a sense of social direction and responsibility; policy issues should be defined so that it is clear if they require research knowledge; research findings should not be counterintuitive, but should be believable on grounds of objectivity and have politically feasible implications; policy makers and researchers should be linked by information specialists capable of translating scientific findings and coupling those findings to policy goals and objectives of the decision makers; value differences among community lobby groups should not be extreme (design-related conflicts should be minimal).

The study of environment policy development relative to environment-behavior issues is a relatively new area of concern for environmental design research. Some research has begun to address the questions of environmental design research utilization at the professional level (Reizenstein, 1975; Cohen, Hetzer, Januk & Tascioglu, 1976; Hiatt, 1979; Cuff & Wittman, 1982) and at the national level (Margulis, 1977; Seidel, 1982).

The study of environmental policy development is an area that received considerable attention in the 1980 EDRA Survey on Research Directions for the Future (Appendix B). A number of policy-making issues have been identified that need attention. One of the greatest needs is an overview of policy-making processes in environmental design—a description of the process, a review of the major theories and research findings, and implications for environmental design research applications. A second need is for additional research on the effectiveness of alternative environmental design policies from users' perspectives as well as from decision makers' objectives. A third area could usefully identify the nature of environmental design research within the decision-making process at national, state, regional, and local levels, as well as in the environmental planning and design professions. How well does the instrumental utilization model proposed apply to the environmental design sciences? How well does this model apply at these other levels of environmental decision making? Can the research findings on factors contributing to successful use of research be applied to environmental design research at the national, regional, or local level? Or, is there a different set of relevant factors influencing the use of environmental design research knowledge in public policy making? Finally, what models are applicable for the private sector?

ENVIRONMENTAL DESIGN METHODS

A major assumption in environmental design research is that the processes of design decision making can be improved through study and refinement. An objective of research on these methods, therefore, is to illuminate the procedures of design and decision making through analysis of the processes, the resultant products, and the weaknesses and strengths of these different methods. Whether the actors are following a logical, rational model for decision making, a not-so-rational intuitive model, or even a political model, they can improve design decisions through greater understanding of the processes they use. Much of the past research on design decision making has dealt with defining and improving the logical self-conscious processes of design professionals. It is interesting to note that the initial formation of EDRA was an outgrowth of the Design Methods Group, and much of the work in this area of research has continued within that group.

Three areas of design methods research pertain especially to the use of environment-behavior research information in the design process: programming, participatory design, and computer-aided design.

Environmental Programming

Environmental or facility programming is the identification of goals, objectives, and values of different users, clients, and designers in the design process (Palmer, 1981). It is the process of identifying problems and the requirement to be met in the development of solutions. The facility programming phases of design continue to show the impact of environmental design research most conspicuously. But the technology of programming is still in its infancy.

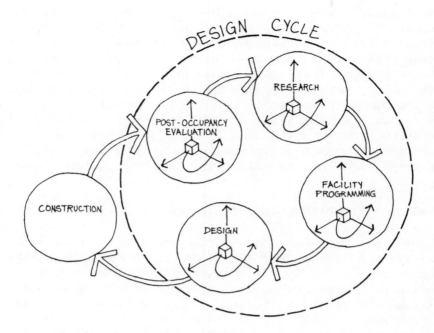

Figure 37 Environmental design cycle.

One of the most recent works summarizing the current state of the art of programming is Palmer, *The Architect's Guide to Facility Programming* (1981). This volume summarizes major writers in the area of programming including: McLaughlin, Pena, Sanoff, White, Preiser, and Davis. Drawing upon their experience, Palmer provides an excellent review of issues and methods of environmental programming.

One of the fundamental needs in this area is to examine the comparative value of different programming methods. Research that describes the pros and cons of the various programming methods is needed and situations where these methods are most appropriate pinpointed. Research in environmental programming could also identify and test techniques for targeting users' needs and attitudes in professional contexts.

Many design professionals and practitioners have complained that much of facility programming has identified irrelevant information which has no direct application in the development of design solutions. Thus further research in programming is needed to develop methods of identifying the critical aspects of user needs which effect design solutions.

Another frequent complaint about programming is that solution criteria often dictate design solutions. Consequently, research is needed on methods of identifying and presenting design criteria and requirements that do not constrain design solutions needlessly.

Finally, research on environmental programming could include the development and test of methods incorporating cost effectiveness, including social cost, into solution criteria. For example, how much can be saved over the life of a building by effective programming? One study that looked at economic incentives of incorporating environment-behavior research in the programming and design of buildings was the recent work of Brill and his associates (Brill, 1982). It can serve as a benchmark for future investigations in this area.

Participatory Design Methods

Another area of needed research is the development of techniques for resolving conflicts between competing user groups as well as between clients and design professionals (Stea, 1967, 1975). Some of the needed research includes: methods for identifying critical aspects of competing goals, objectives, and values of different actors; and methods for developing and testing techniques for resolving those conflicts between competing interest groups.

Critical to resolving conflicts is identifying trade-offs and the cost and benefits, both financial and social, of alternative procedures and solutions within the design process. There is a need to develop predesign evaluation techniques to examine alternative design solutions in behavioral terms. There is also the need for developing methods of projecting probable future states of the environment and cultural institutions, along with methods of predicting impacts and consequences of alternative solutions.

Finally, inquiry into participatory methods of design, decision making and research should include techniques for enhancing the use of interdisciplinary collaboration (Zeisel, 1981). Although there has been a great deal of attention to the need for interdisciplinary collaboration in design and research efforts, little is known about the depth of cooperation, or what the impacts may be. It has been assumed that research and design collaboration will produce better results. Research is needed that evaluates the benefits of interdisciplinary collaboration in both design and research efforts. Where is collaboration being done at present? Who is collaborating with whom? How effective is collaboration versus noncollaboration in design and environmental research? Are the benefits worth the cost for collaboration? If collaboration does provide positive results in design, research, and applications, how can further collaboration be improved and promoted?

Computer-Aided Design

Other issues important for the next decade include research on and applications of computer-aided design systems and ways to make them more responsive to environment-behavior knowledge (Eastman, 1975; Mitchell, 1977; Gero, 1977). Industrial nations are entering a new era, one of a highly sophisticated, information-rich, and computer-aided society. In design decision making it is imperative that new information systems incorporate human, behaviorally based data. Further research is needed in the areas of computer-aided design methods that can be accessible conveniently to design professionals and environmental policy makers. In addition, further research is needed in the area of artificial intelligence and simulations of environment-behavior relationships for modeling human behavior and the environment.

ENVIRONMENTAL EVALUATION

Environmental or post-occupancy evaluation (POE) is the evaluation of an environment after it has been occupied. A POE can be designed to study specific issues within a building or setting, or it can be designed to evaluate a project across a number of theoretical issues to determine its effectiveness for human users (Reizenstein & Zimring, 1980). A POE can evaluate a setting's effectiveness in economic and technical terms as well as socio-behavioral terms (Rabinowitz, 1979).

A POE generally is descriptive in nature, using observation and questionnaires as research tools for study, and is most often aimed at application. This focus of descriptive study and applied orientation distinguishes a POE from more basic environment-behavior research. Thus POEs have been defined more specifically within three conceptual dimensions: generality, breadth of focus, and applicability (Reizenstein & Zimring, 1980).

Other ways of understanding the nature of a POE is whether or not it is a cross-sectional or a longitudinal study, and whether it involves single versus multiple settings. POEs tend to exist at one end of the continuum—specifically, one of a kind studies of particular settings. At the other extreme are the more costly cross-setting studies characteristic of basic environment-behavior research.

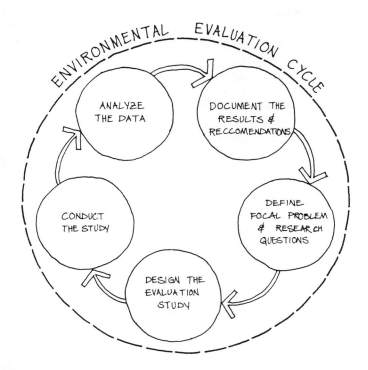

Figure 38 Environmental evaluation cycle.

Within this framework, two major questions needing further attention are how can building evaluations be included in the normal cycle of programming, design, construction, and evaluation; and how can the POE process be improved and incorporated into longitudinal research investigations? There is also a need for the development of a systematic way of developing POE research questions, determining variables, instrumentation and data handling that can be consistent from one POE to another. More general findings can be drawn from the disparate data contributing to the development and testing of environment-behavior theory.

Marans (1984) has identified other priority research needs in the evaluation area (compare Moore & Marans, 1982): ways to specify measures of success; calibrate the full range of environmental variables which could be linked to outcome variables; improve research designs (quasi-experimental rather than one-shot case studies); incorporate multiple settings and longitudinal design, and use new methods for analyzing data rigorously. To advance the state of the art of environmental and building evaluations, Marans recommends that the field turn its attention to the larger field of program evaluation research (which has its own journals, newsletters, formal organizations, and lobbying efforts).

INQUIRY INTO ENVIRONMENTAL DESIGN RESEARCH METHODS

Preceding sections have focused on the research needs of different phases of the environmental policy-making and design cycle. In this section, we focus our attention on improving the methods and techniques of research itself.

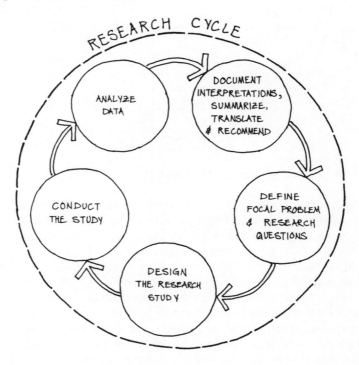

Figure 39 Environmental design research cycle.

Not only does the field borrow theory, models, and concepts from related social sciences, but also research tools and techniques. The area of environmental design research uses traditional methods of scientific research, but it is not wholly defined by them. It uses precision and quantification as well as qualitative methods.

The purpose of these methods and instruments is to define and measure the variables of the environment and of behavior in qualitative and quantitative terms. Measurements of the relationship between people and their environment allow assessments of the quality of life to be made. The charting and understanding of the relationship between objective changes and subjective responses, and development of reliable indicators by which we might gauge advances in social well-being, are still relatively undeveloped (President's Commission for a National Agenda for the Eighties, 1980).

Some major works in the field of environment and behavior research that describe research include Michelson's *Behavioral Research Methods in Environmental Design* (1975) and Zeisel's *Inquiry by Design* (1981). These volumes provide excellent overviews on research instruments. One challenge is to decide which instruments are most appropriate in a given situation and to define which instruments best match research questions being asked (Patton, 1980). Another priority need is an examination of the appropriateness of different research designs for the field, from exploratory case studies and observational designs to quasi-experimentation and experimental designs (Campbell & Stanley, 1963; Cook & Campbell, 1979).

Another area of research methodology that needs clarification is the techniques of scholarly and historical research, and the qualitative approaches of anthropology. Foster (1969) includes the techniques of case studies, comparative case studies, and participant observation (see also Webb, Campbell, Schwartz & Sechrest, 1966; Patton, 1980; Brewer & Collins, 1981).

There is also a need for development of measurement capabilities that will permit consequences of design, management, and policy decisions to be formulated within a comprehensive and internally consistent framework. Among the professional disciplines involved with the habitability of places, there are no commonly shared criteria for assessing the direct or indirect consequences of the actions of decision makers. The development of a dialogue between disciplines is made more difficult by differences in the procedures and processes by which professions communicate and operate and priorities to which each responds. In order to be understood by all concerned, measurement systems need to be developed which permit the decision maker creatively and consciously to manipulate the variable under control. The researcher should be able to identify and manipulate the variables related to the consequences of design decisions in a precise and objective manner. Research methods and measures are thus needed that are shared by a significantly broad professional constituency in design, management, and policy decision making. In addition, precise and objective measurement capabilities sensitive to environmental and behavioral variability are needed.

Finally, approaches to measuring relationships between the organization of the environment and patterns of individual and collective behavior are needed to measure both direct and indirect consequences of decision making, and behavioral and environmental attributes about which decisions are made.

RESEARCH ON IMPLEMENTATION STRATEGIES AND KNOWLEDGE UTILIZATION

Additional research on application methods consists of studying the relationships or connections between the various stages in the environmental policy-planning and design process sketched earlier. Priority areas of applications research include: study of data management and translation; strategies for disseminating research findings to design and policy-making professionals; research on implementation of environment-behavior research findings; and the study of effective media, techniques, and tools of communication.

It is important for the vitality of environmental design research that all actors involved in decision making, the design and building industry, as well as policy making and planning, continue to be involved in the research side of the field. It is equally important for researchers to be involved in the design/build side. Researchable issues should inspire collaboration between the environmental research community and designers, planners, and decision makers acting on the physical environment.

Research on mechanisms whereby research findings are actively disseminated to practitioners and policy makers is urgently needed. First, a comparative analysis of various strategies for managing research information prior to communication is needed. These management systems would include information data banks such as the IBIS system (Issue Based Information System; Rittle & Webber, 1973), the ASMER system (Associate for the Study of Man-Environmental Relations; Murtha, 1978); or the TEAG system (The Environmental Analysis Group; Szigeti & Davis, 1980). With recent advances in information-processing systems now widely available, the field may be at the stage of only needing application of current

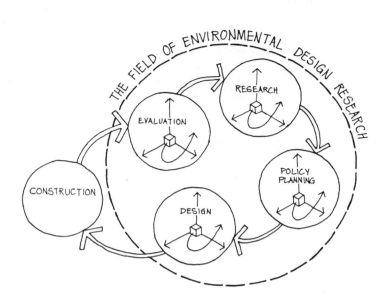

Figure 40 The field of environmental design research (environment–behavior research and design applications).

technology. Further research may identify ways of encouraging the use of these systems. Computerized access to cataloging research reports may not be as effective as direct access to research results with specific design recommendations for specific problem types, issues, or place settings (that is, a computerized design guide data base and information system).

It seems apparent that design guides, generalized design programs, and design criteria based on behavioral data have become the most popular technique for collecting and disseminating environmental design research for specific building types or design problems. Yet, how well do design guides communicate? Do they speak to the images with which designers work? Are design guides the most effective means for disseminating behavioral research implications? Are, in fact, design guides being used actively in the design process?

Research is needed to identify and develop more effective media for communication, for example: research to identify media and communication systems that best encourage the use of environment-behavior research findings; research on information and dissemination methods that reach the greatest audience; and research on information and dissemination methods that have the greatest impact in changing policy, decision-making, and design procedures.

Much of the dialogue on research dissemination has been directed toward designers and other decision makers. More effort is needed at informing the users, the clients, and the ultimate recipient of behaviorally based design solutions, the consumer. The designer and decision maker may be much more responsive to environmental design research if the client and public are clamoring for its use and application.

10

CONTEXTUAL ISSUES: ENVIRONMENTAL PROBLEMS AND CONSTRAINTS

Some of the larger social and environmental issues that provide the context for research and professional practice in this field and which serve to give it impetus and tend to constrain its development. Included are discussions of issues of limited resources (energy, food, production), deterioration of the environment (environmental pollution, and the improvement and reuse of the infrastructure of our environment), social change (the coming of the post-industrial society, geographic mobility, lifestyle, technology), and the changing economic and political climate (allocation of resources, the organizational arrangements for resource utilization).

I do not wish to be overdramatic, but I can only conclude from the information that is available to me as General Secretary, that the members of the United Nations have perhaps ten years left in which to subordinate their ancient quarrels and launch a global partnership to curb the arms race, to improve the human environment, to defuse the population explosion, and to supply the required momentum to develop efforts. If such a global partnership is not forged within the next decade, then I very much fear that the problems I have mentioned will have reached such staggering proportions that they will be beyond our capacity to control.

U Thant, 1969

So what else is new?

Michael Brill, 1982

There are a number of issues or themes that are important to environmental design research because of particular circumstances facing the world. These contextual issues are defined by the critical nature of social, political, economic, cultural as well as physical environmental trends facing industrialized and developing nations.

109

These trends are not to be considered as speculations about the future. Rather, contextual circumstances and trends exist now and will continue to affect the world in the future. Because the nature of environmental design research is applied and problem-oriented, discussion of these trends is critical to the development of any report on future research needs. These overriding trends shape the context for environmental design research.

These contextual themes are discussed in the following areas: pressures due to continued limited resources; continued deterioration and pollution of the environment; technological change; the coming of the postindustrial society; and impacts of the changing economic and political context. A concept that weaves its way through all these issues is the notion of change itself, how change comes about, and what the effects of time and change are on environment-behavior relationships.

LIMITED RESOURCES

Limited global resources is one of the critical issues facing the world. It includes limited energy resources, as well as limits to food and industrial production.

Limited Energy Resources

The impacts of energy shortages are far-reaching and severe, impacting the quality of life for all nations. Dwindling fossil fuel supplies endanger economic health and the quality of life of all nations. During the years following the OPEC oil embargo, the United States devoted much human energy to reducing fossil fuel consumption. Because approximately one-third of U.S. energy supplies are used to heat and cool buildings, the conservation of energy in building design and management has become a critical planning and research issue in the 1980s.

Often the technological approach dominating energy research in the 1970s has taken the viewpoint that the solution lies in mechanical efficiency and, more recently, expanding the search for the remaining limited fossil fuel sources. From this viewpoint, buildings and machines are assumed to use energy because machines and buildings are inefficient and because management of those machines and buildings is inefficient.

The environmental design research approach to energy and the environment has proceeds from a different set of assumptions—

buildings use energy because of people, not machines. The relationship between people and the environment and the need for comfort are critical elements of the environmental design research approach to energy and may yield a substantially different set of answers. Making machines more efficient is useful and shouldn't be excluded, but the problem also involves how to provide for human satisfaction, fulfillment, comfort, and well-being while using a minimum amount of limited nonrenewable fuel sources. One of the landmark works describing relationships between energy use and human behavior is Odum's *Environment, Power, and Society* (1971).

Limited Food and Production Resources

In many third world nations there is a shortage of fuel for cooking and heating. Many people have turned to animal dung for use as fuel. While not a new occurrence, this practice reduces the source of fertilizers and in turn food production. The end effect is a spiraling decline in the standard of living and the quality of life. Research could valuably address the questions of resource alternatives and ways of interrupting or ameliorating this environment-behavior spiral in developing nations.

The relationship between people, environment, and human need for physical as well as psychological satisfaction, personal fulfillment, comfort, and well-being is a critical element of the environment-behavior research approach to all problems of limited resources. A better understanding is needed of the relationships of people to the designed environment as a component of the larger ecosystem. The intent of such research is to develop an infrastructure and cultural systems appropriate to the world's limited resources. This conscious connection between elements of the social and natural environments is critical if we are going to continue to survive and improve the quality of life.

DETERIORATION OF THE ENVIRONMENT

The continued deterioration of the environment is a second major contextual issue around future research. Trends within the context of a deteriorating environment include: continued deterioration of the existing urban infrastructure, and increased pollution of the ecosystem, water, air, and land.

Improvement and Reuse of Infrastructures

Due to the depletion of resources for new construction and development, it will be critical to improve and maintain existing structures in the built environment to meet changing needs of society and social institutions. In many societies, maintenance and preservation of existing places is tightly woven into cultural values of those societies. Older places are held in high regard for their historic value. Many older buildings are preserved and retrofitted for continual use. In the United States, however, historic preservation is a relatively new phenomenon (*Newsweek*, 1981). This is partly due to the newness of the environmental settings and partly due to the social value of always having the newest product on the market.

Research is needed on the implications of the reuse of existing buildings to meet the needs of a changing society. To what extent can improvements in the existing structure of the built environment be changed, reused, or improved to satisfy the needs of people? Conversely, what stresses occur in social structure and in individuals if people or the environment are not able to adapt?

Environmental Pollution

Environmental pollution is a continual threat to the quality of life for all nations. As with much of the research on energy consumption, pollution is often seen as the result of technical and mechanical problems. The environmental design research approach to pollution proceeds from the assumption that people and people's behavior contribute to pollution. The relationships between people and the pollution of the environment are critical issues for further research. Research is needed to identify the relationships between existing cultural and behavioral patterns that lead to deterioration of the physical environment. Theories and models for explaining and predicting interactions of the complex system are needed.

SOCIAL CHANGE: THE COMING POSTINDUSTRIAL SOCIETY

Much of environmental design research is static in nature, developing insights on environment and behavior relationships within a set social context. These limitations are acceptable if one is to gain an understanding of the current cause-effect relationships of the environment-behavior interface. However, these insights and implications can become even more complex and powerful when seen within the context of social, political, and environmental changes. For example, the relationship between human behavior and the socio-physical environment may be more complex when understood within the changing context of a highly sophisticated, technologically advanced, and information-rich future of the Western world.

Research within the context of rapid social and technological change includes: developing an understanding of geographic mobility; major changes in lifestyles, sex roles, and family structures; the growth and impact of an information-rich society; reindustrialization of the Western world; technological growth; and industrialization of the third and fourth worlds.

Geographic Mobility

For years planners have been aware that social patterns and customs change as a result of shifts in populations. In the United States, population shifts from the industrial northeast to the sunbelt are of current interest to demographers (*New York Times*, 1981; *Wall Street Journal*, 1980; compare Stokols & Shumaker, 1981). What shift in lifestyles can be predicted from these mass population shifts? Can these changes be controlled, directed, or even, should they be?

Population shifts have long been occurring in developing countries. In many, massive migrations are occurring from the agricultural areas to the industrialized cities. These population shifts to the cities parallel the population shifts in the technologically advanced, industrialized nations during the 19th century. Mexico City has one of the fastest growing populations in the world. If trends continue as they are now, Mexico City will soon become the largest city in the world. Much of this growth has occurred in squatter settlements which have overwhelmed planners and environmental policy makers with problems of health, welfare, transportation, housing, and education, as well as the economic problem of providing jobs.

Research issues in environment-behavior relationships may help to improve such conditions. What changes must occur in the physical environment as well as in forms of social, political, and economic conditions to effectively change and improve the living conditions of so many people? And, most importantly, can these changes occur without completely disrupting people's lives through physical revolution or conflict? Can these changes occur in a peaceful way before the complete deterioration of the environment? Can alternative forms of settlement provide answers to these questions?

Changes in Lifestyles

Societies are not static. Values, social roles, family structures, and cultural institutions all change over time. The relationships between people and their environment will be affected by these changes. For example, the changing roles of women and men in society has been an area of great interest and concern since the mid-1970s (see Chapter 7). It has also been widely speculated that the coming of the postindustrial society will provide greater leisure time activity. This is partly a reflection of the traditional work ethic. But we must not ignore the importance of leisure issues and their relationship to the quality of life and the enviroment.

More research is needed on the effects of changing lifestyles, social roles, and institutional structures on the built and natural environments (Hunt, Feldt, Marans, Pastalan & Vakalo, 1984; Feldt, Marans & Pastalan, 1983). Can a theory of social change and the environment be developed to explain the nature of this phenomenon? What are the effects in terms of neighborhood patterns and housing due to changing social roles? Is there a pattern to these changes related to environmental conditions?

Technological Change

It has been said that the technologically advanced nations are entering a new era in history. This new twenty-first century society has been popularized as an information-rich, postindustrial society (see Kahn & Bruce-Briggs, 1972; Bell, 1973; Toffler, 1980). This new era promises to bring forth advances in communication, technology, health, and energy which will revolutionize culture as we now know it. The implications are not limited to the design of new environments, nor understanding changes in society and social institutions. There is the real possibility through genetic engineering to change the basic matter of life itself. These dramatic changes present a wealth of urgent questions for environmental design research.

Environment-behavior research needs to focus on developing an understanding of how new technologies affect social organizations and institutions and the resultant effects on the built environment. Two changes of interest are developments in information and communication technologies and the changes in the industrial character of the work place.

Advances in the use of computers and information systems have already affected the way people live (cf. Becker, 1981). In the Western world, home computers and information systems are becoming commonplace. How will this proliferation of information effect privacy, family structure, the home, and the work place? What new kinds of environments must be designed to cope with these changes? Will humanity be able to cope with an overabundance of information? Or, will societies continue to develop machines to act as intermediaries between information and people?

Reindustrialization and the Changing Work Place

Because of advances in technology, many futurists have described a new work place, and a new industry based not on production but on information-processing services. Over time, more of the work force will shift from basic production to information processing and service jobs. The industry that remains is expected to be taken over by highly computerized robot systems. The present industrial work places of the United States would be reindustrialized to stay in step with the rest of the highly developed world.

As described by Kahn and Bruce-Briggs,

> This does not imply that industry will decline in real terms (although it might) and it certainly does not imply that industry will not play a vital role in the society. This argument is that industry, that is, secondary industry (manufacturing) and its direct services will go the way of primary industry (farming, fishing and mining) and its direct services. Instead of having ninety-five percent of the labor force engaged in primary industries, in the United States today less than five percent manage to do these things for us. A similar shift from secondary to 'tertiary industries' (services) is taking place today. We know of no one who doubts these trends will continue (Kahn & Bruce-Briggs, 1972).

As this reindustrialization occurs, how will it affect the worker? What new environments will be needed for a computerized, service, and information-oriented society?

Much of the developing third world is facing similar processes of change. While a number of the Western, technologically advanced societies are moving from an industrial to a postindustrial society, third world countries are moving from agrarian societies to advanced technological societies. The impacts on human development and the form of the environment are immense. Many countries have not been able to adjust to such dramatic environmental changes without overwhelming social conflict.

These social change issues are not limited to questions concerning geographic mobility, changes in lifestyles, technology, and industry. Changes are occurring that effect every aspect of life and culture: in economy, religion, education, health, and social values. each of these aspects of a changing society can be studied for their implications for environment-behavior-design relationships.

THE CHANGING POLITICAL AND ECONOMIC CONTEXT: THE STUDY OF ENVIRONMENTAL INSTITUTIONS[1]

The differences and changes in political and economic structures have an enormous impact on the attitude with which planners, designers, builders, and users approach the design and construction of the environment.

One thing the past 25 years of environmental design research has taught us is that developing an identifiable field of theory and knowledge does not by itself assure that the knowledge gained will be used by the building community, or by the schools that train future members of the environmental professions. The difficulties involved go beyond the weaknesses of the methods that designers use for increasing the research-based content of their work. They also depend on the structure of political and economic relations through which decisions are formulated, and the importance that a culture assigns to environmental quality.

The political and economic context for environment design research has enormous impact on the kinds of research issues that are supported, where the support comes from, the nature of funding, and the application of knowledge. It is critical that research be conducted on understanding political and economic institutions, and on developing new forms of research for use in this context.

Allocating Resources

The construction of built environments demands large inputs of material, labor, and financial and intellectual capital. These resources are limited in any culture, and therefore, an interesting question arises: How are the decisions made in allocating resources, say, to housing rather than to factories, or highways rather than railroads? Recently the argument has been made that some of the problems of the United States' economy are the result of too much investment capital flowing in the last 50 years to housing, and not enough to new factories and industrial equipment. How did this happen? If resources begin to shift to industrial plants and away from residential environments, which economic and social groups will bear the burden of this reallocation? The outcome of these policy debates will influence the form of the environment for decades to come. More needs to be known about the criteria and institutional arrangements for allocating resources and their impacts on the form, shape, and nature of the physical environment.

The Organizational Arrangements for Utilizing Resources

Regardless of which type of environment a society constructs, the process of making use of resources can be more or less efficient. A standard criticism about the building industry in the United States is that it departs from the production model of advanced industrial societies. The building industry (and this includes the design professions as well as construction firms), critics claim, is much too fragmented, relies on craft methods of production, and is often unable to regulate its market. Some scholars feel that the inefficiency is intrinsic to the type of commodity produced; others take the view that much greater rationalization is possible, and point to the lower cost generated by economies of scale and industrial production methods. This debate and its resolution have consequences for environmental design research, because greater efficiency in production has generally enabled industries to spend more income on research and development activity. Further research on the development and building processes of the construction industry may provide interesting solutions to improving the use of resources.

Institutional Obstacles to the Utilization of Advanced Technology

Along with the complaint about fragmentation, the building industry and the environmental professions have been accused of resisting technological innovation. Reports have documented the effort of companies that manufacture building materials to subvert government programs intended to develop and introduce new materials and new manufacturing techniques (Gutman, 1981). Many local building officials and small design firms have joined in the subversive tactics. We should know more about this area of scholarship, again, for the reason that opposition to technological innovation in the engineering and materials areas usually extends to all forms of new ideas and information, including the knowledge and skill associated with environmental design research. In this case, the issue becomes how to restructure the environmental institutions to encourage greater use of research-based knowledge. This is absolutely essential if the field is to make genuine progress in dealing with the perennial concern about information dissemination. The quality of output is not the obstacle; the organization of the building industry is where the problem lies.

Consumer Values and Resistance to Advanced Planning and Building Concepts

Those working in environmental design research sometimes exhibit ambivalent attitudes toward the consumer. In some situations, the consumer is hero and the field is protector. This is particularly so in those cases dealing with clients or developers who, in their pursuit of a quick solution and a hefty profit, ignore the needs of the worker, the poor, the sick, and the disabled. But in other situations, the consumer becomes part of the problem, for example, when the consumer is not aware of the benefits of public transportation or compact housing development or flexible work spaces. The confusion about whose side to take illustrates the need for environmental design research to recognize that the consumer is just another actor, or class of actors, within the total environmental institution. The field may adopt a critical objective attitude toward the consumer's requirements, and not assume that in every situation the consumer necessarily represents the public interest as a whole. One way to achieve this perspective is for environmental design research to develop a research interest in the entire structure of environmental decision making, and to study the make-up of the class or stratification system that is included in the set of institutional arrangements that regulate the environment.

Values, Attitudes, Social Roles, and Organization of the Environmental Professions

One of the classes that make up the stratification system of the environmental institutions is the environmental professionals: industrial designers, interior designers, architects, landscape architects, urban designers, urban planners, and environmental policy makers. This is an especially critical group to know more about because environmental design research shares a social role with them as experts in the division of labor that governs environmental decision making. It is important that we should be familiar with the ideology of the design professions, the directions in which they are moving, both in terms of beliefs and internal organization. At the moment, the design professions are probably moving toward a greater concern for the social effects of design and in the direction of rationalizing their entrepreneurial operations. The combination of these trends is likely to increase the market for environmental design research. It may also influence the kinds of research questions that environmental design research should be addressing. Professional associations (AIA, APA, ASLA) are very interested in finding a group that can supply information about political and economic trends that impact their market position (see American Institute of Architects, 1982). Among the questions that interest them is the shift taking place in the public's image of the function of the different professions, changes in effective demand for various building types, and the emergence of new building types and user requirements.

The addition of research to environmental institutions means that the field will have to include political science and economics under the environmental design research umbrella. Also, it suggests the need for adopting a comparative, or cross-national perspective to all of environmental design research.

NOTE

1. This section is based heavily on a position paper prepared by Robert Gutman, a Task Force member and Chair of the EDRA Board of Directors, 1981-1982.

PART III
RESEARCH
IMPLEMENTATION

11

STRATEGIES FOR IMPROVING RESEARCH

Available and possible strategies for further improving environmental design research and applications through increased federal and private support, improved organizational and political climate for research, and improved graduate education and public environmental education. Six major objectives are identified, under which are discussed example implementation strategies. The objectives include the identification of major problems requiring environmental design research, increased public awareness of the needs and utility of environmental design research, increased impact on legislation affecting the climate for research, increased influence on the patterns of funding from both the public and private sectors, improvements in graduate education in the social sciences and the environmental professions, and the initiation of programs and processes within scholarly and professional associations.

We believe that scientific knowledge provides an essential foundation for modern society and that the preservation and further progress of the society calls for the vigorous pursuit of science, both basic and applied.

Herbert Simon
Social and Behavioral Science
Programs in the National Science Foundation, 1976

The authors would like to thank the following individuals who contributed ideas for this chapter: Roberta Miller of the Consortium of Social Science Associations (COSSA) whose talk at EDRA 13 led to many of these ideas; John Archea, Robert Marans, Robert Shibley, Francis Ventre, and Ervin Zube who contributed many more of the ideas during a meeting at EDRA 13, and especially Francis Ventre who has reviewed the chapter carefully and made many valuable suggestions. The chapter is also based on a content analysis of the responses to the EDRA survey (May 1982).

This chapter presents strategies to furthering knowledge of environments, their design, and their effects on people by encouraging more and higher quality environment-behavior research studies and applications.

How much direction should this or any field of research be given? Scientists in and outside the field have expressed concern about possible conflicts between innovative research and attempts to direct it. Two articles in *Science* (Muller, 1980; MacLane, 1980) warn that well-meaning attempts to force science into preconceived directions can be counterproductive if these attempts are not carefully balanced with an appreciation of scientists' expertise in knowing what lines of inquiry are most worth pursuing. They also warn that some recent developments in federal funding patterns, and especially in reporting requirements, are having a serious impact on research in universities, government, and the not-for-profit private sector. U.S. Senator Moynihan (1980) raises grave concerns about major transformations and regulations of the research community brought about by shifts in government policy toward research. Concerns like these may have led scientific scholars and commentators, like Simon (1976), to speak out on behalf not only of applied research but also of basic research that leads to fundamental advances in knowledge and resulting important practical applications.

GOALS

The purpose of environmental design research is to advance the field of environment-behavior studies in both scientific and professional forms. It advances scientific inquiry into environment-behavior relations, and it promotes the use of research findings to promote human welfare and the quality of life through improvements in the quality of the physical environment.

The environmental design research field can influence the pattern, standards, and timing of research and applications in many ways, both direct and indirect. These include establishing research needs and priorities and influencing the quality of research through use of incentives, general information, and persuasion techniques.

OBJECTIVES AND LEVELS OF IMPROVEMENT

To achieve this general goal, this chapter presents six objectives:[1]

1. To identify major problems requiring environmental design research.
2. To increase awareness of the needs for and utility of environmental design research.

3. To influence legislation affecting the climate for research.

4. To influence patterns of funding for environmental design research from both public and private sectors.

5. To increase and improve graduate education in environmental design research (environment-behavior studies and applications) and its sub-areas in the social and behavioral sciences and the environmental professions.

6. To invigorate existing programs and processes within scholarly and professional associations, and to initiate new programs where needed.

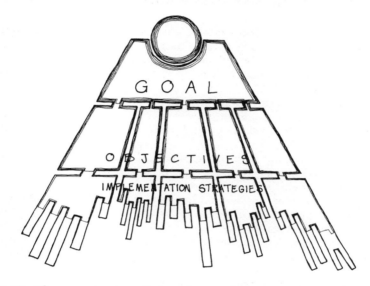

Figure 41 The organization of objectives and strategies for implementing programs of environmental design research.

THE IDENTIFICATION OF PROBLEMS REQUIRING ENVIRONMENTAL DESIGN RESEARCH

As a part of developing this book, the study team surveyed the environmental design research field (researchers, practitioners, and government officials) to identify problems requiring environmental design research. A number of important problems were raised, and these are reflected throughout this book and its appendixes. Problems identified by environmental professionals (architects, planners, interior designers) tend to be practical problems about the quality of the environment for which they, in their professional roles, have inadequate knowledge on which to base decisions. The study team also sought other reviews of the research literature, especially those describing recent developments in substantive areas and highlighting problems needing attention. Most tended to be theoretical problems suggested by the nature of the field and its current scientific status.

This first objective, then, is to advance the state of research on environment-behavior relations and applications by continuous identification of problems—both practical and theoretical—whose effective resolution requires environmental design research. To implement this objective, several strategies are recommended:[2]

1. Identify the demand base for environmental design research: who needs it, who wants it, who uses it. As part of this strategy, draw practitioners closer into the research community, and vice versa, so lines of influence and information can flow more freely and quickly between research and utilization.

2. Conduct a broad identification of research problems from the perspective of the environmental professions. Support a series of probability samples of all the environmental professions about their information needs each five years.

3. Meet biannually with the research-oriented standing committees of the several professional associations (e.g., AIA, ASIA) and technical groups (e.g., ASTM) to identify their research needs and priorities.

4. Encourage the periodic and systematic review of the literature on research and applications in various substantive topic areas (e.g., those suggested by the chapter titles and sub-headings of this book) to identify recent developments, problems needing resolution, and priority research issues. This could be done by initiating a series of volumes to update advances in the field every 15 to 18 months (e.g., the new EDRA sponsored Advances in Environment, Behavior, and Design series).

INCREASING AWARENESS OF THE NEEDS AND UTILITY OF ENVIRONMENTAL DESIGN RESEARCH

Public decision makers at all levels of government, and professional decision makers at all scales of the environmental professions need to be made more aware of the needs for good environmental design research and of the usefulness of such research. This objective can be achieved in part by implementing the following four sets of recommendations:

1. Evaluate and communicate examples of effective environmental design research:
   a. Diagnose examples of the effective utilization of environmental design research and identify where environmental design research has had important impacts on action programs (e.g., Wener, 1982).

b. Analyze examples of the cost effectiveness of environmental design research, where the costs of not doing research outweigh the added product, personnel, or process costs of not knowing (e.g., Brill, 1982).

c. Develop and communicate ways in which organizations can rethink and recalculate their own in-house costs and the added value of using research.

d. Document and communicate research-based innovations.

e. Initiate a major research utilization demonstration project, evaluate it, and communicate the results widely.

f. Encourage other forms of identification and communication of positive examples of environmental design research use at all scales and levels.

2. Initiate media campaigns to inform public and professional communities of the need for and utility of environmental design research:

a. Develop a media film that shows environmental design research as a new field of studies with recognized scientific status and practical utility.

b. Develop slides, tapes, books, manuals, and fliers for dissemination to professional and government bodies and the public.

c. Explore the possibilities of an educational television or national public radio show on environmental design research and applications (e.g., a forthcoming Wisconsin Public Radio series).

d. Develop an informational newsletter for major federal, state, and local agencies, and for professional associations to keep them abreast of developments in the field.

3. Develop an active and regular dissemination program to communicate research findings to the environmental professional community:

a. Develop a monograph series on advances in environmental design research and applications (e.g., the new EDRA-sponsored monographs to be published by Praeger Publishers).

b. Seek funding from federal agencies or private foundations for participation in current efforts (e.g., at the National Building Museum) to establish an information storage, retrieval, and dissemination system, and for regular updates of the data in that system.

c. Develop a series of other hard-copy books, monographs, and reports based on data in the system, that is, compilation and translation documents on particular place types, environmental user groups, and socio-behavioral phenomena.

d. Study other successful professional communication groups (e.g., the American Planning Association's Office of Research) and other information storage and retrieval systems (e.g., IBIS and others mentioned in Chapter 9) for utilization in the field.

e. Seek endorsements of all EDRA documents and publications or joint publications with other associations concerned with research on aspects of the human environment.

4. Increase recognition, incentives, and public announcement of successful environmental design research:

a. Offer annual awards for members of the field who have achieved the objectives of environmental design research suggested in this book.
b. Identify who will gain by personally announcing research grant awards or research honors (e.g., local congressional representatives).
c. Encourage the general public and elected officials to advocate for increased support for environmental design research.
d. Develop political and educational activities on behalf of environmental design research.
e. Increase incentives for conducting interdisciplinary, environmental design research in universities and private practice, including public recognition for excellence in conducting research or in using research in environmental applications (e.g., encouraging state legislators to give state citations for research and applications as has been done in Wisconsin).
f. Seek an immediate voice in the research awards of those disciplines that comprise environmental design research (e.g., the American Institute of Architects' Research Medal, Progressive Architecture's Design Research Award, the American Society of Landscape Architects' Bradford Williams Medal, the American Psychological Association's Distinguished Scientific Career Award and Early Career Award, and the American Sociological Association's Sorokin Award and Stouffer Award, etc.).

INFLUENCING LEGISLATION IMPACTING ON THE CLIMATE FOR ENVIRONMENTAL DESIGN RESEARCH

Several commentators have encouraged the environmental design research field to actively work to influence legislation and other regulations and standards (especially at the federal level) that will create a climate more supportive of environmental design research. Three recent examples of types of legislation and standards in which the Association has taken an active interest are: U.S. Senate Bill 5090 (the Moynihan Bill) which advocated evaluation research on all projects supported with federal money; the U.S. Department of Housing and Urban Development's housing evaluation research policies and minimum property standards; and the U.S. Department of the Army's research-based technical manuals and design guides for construction in family housing areas at military installations worldwide. Most of the ways in which the field can impact legislation and regulations affecting research depend on legislative initiative; three particular strategies are recommended for impact on this process:

1. Monitor legislative programs of interest to the environmental design research community, and make this information widely known in the research community.

2. Draft model legislation at the federal as well as local level, that encourages the development and use of environmental design research knowledge, e.g., legislation encouraging facility programming (as has been done in Canada), post-occupancy evaluation, etc.

3. Actively support bills of interest to the field of environmental design research when they are in the process of development, in committee, and in Congress (and, analogously, support state and local legislation, and legislation in other countries when it helps to develop a climate supportive of environmental design research).

INFLUENCING PATTERNS OF FUNDING FOR ENVIRONMENTAL DESIGN RESEARCH

A central purpose of this study is to influence funding in the public and private sectors for environmental design research and, in particular, for the priority areas identified in the document. One of the prime audiences for the document, therefore, is program directors and their superiors, review boards, and directors in the nation's research agencies and private foundations.

Most research in the field of environmental design research has been supported by five public agencies: the National Endowment for the Arts; the National Science Foundation; the U.S. Army Corps of Engineers; the U.S. Department of Housing and Urban Development; and the National Institute of Mental Health. Programs in some of these agencies have been seriously cut back or eliminated in recent years, especially in interdisciplinary and new areas of investigation (Simon, 1976). While the National Endowment for the Arts Design Research Program has expanded, and the National Science Foundation has begun a new program, environmental design research's other major nondefense-affiliated federal research organization, the National Bureau of Standards, Environmental Design Research Division, was abolished in 1981. With the increasing pressures on limited resources in the nation's universities, and with increasing pressures for immediate applications from the private sector, it is imperative that federal, state, and local governments remain actively involved in the funding of environmental design research. It is also imperative that the field become actively involved in seeking other avenues of private sector funding for research. Three recommendations for helping influence patterns of funding in the public and private sectors are:

1. Identify the major potential funding sources for environmental design research:
    a. Identify and work with the appropriate levels of federal agencies regarding environmental design research.
    b. Identify the major private sector sources, e.g., private foundations, industry, major corporations.
    c. Match various interests with potential sources.
    d. Develop a data bank of sources and make them available to all members.
2. Understand the political process at the federal level of decision making surrounding research, and work actively with other organizations to increase and direct federal funding within this understanding:
    a. Analyze political processes and explore the feasibility of setting up various alternative paths for affecting the political system.
    b. Map out the structure of opportunities for each different agency, committee, or other body identified in the study.
    c. Focus on incremental changes to existing programs in the direction that the programs are changing.
    d. Prepare reports arguing that research support should be bipartisan.
    e. Band together with other associations and interest groups with similar concerns about research on the human environment (e.g., the former Task Force on Human-Environment Research and Applications; see Appendix B for its organizational and individual members).
    f. Develop a multi-disciplinary direct action lobbying group.
    g. Lobby at appropriate levels within agencies and Congress.
    h. Give attention to particular groups of congressional representatives and senators who chair important science and environmental committees and who are influential on critical issues.
    i. Focus on the budgets of one agency at a time.
    j. Identify members of universities who are on science policy or review boards and influence them relative to the needs and benefits of environmental design research.
    k. Work with larger professional organizations to achieve leverage to get public policies changed (e.g., AIA, APA, ASCE; cf. AIA, 1982).
3. Work together with other associations in environment-behavior research and applications to identify and encourage programs of research that will make a difference:
    a. Encourage support for larger programs of research.
    b. Encourage long-term support for successful programs of research.
    c. Encourage capital funds for equipment.
    d. Identify and support research that can change public opinion.
    e. Support research that is of high priority to the future development of the field, as identified in this and subsequent documents (e.g., theory development and testing, conceptual research, place research, user group research, socio-behavioral phenomena research, process research, and contextual research).

f. Support research that has a high probability of leading to resolution of important issues facing the field.
g. Support research that has a high probability of contributing to the conceptual and methodological sophistication of the field.
h. Support research that has a high probability of contributing to the solution of major human-environmental problems.

THE IMPROVEMENT OF GRADUATE EDUCATION IN ENVIRONMENTAL DESIGN RESEARCH

The long-range success of the nation's efforts to improve the quality of research bearing on human-environmental decision making lies in the education of the next generations of scientists and practitioners entering the field, and the continued training of those already in the field. This objective can be accomplished on three levels:

1. Encourage the development of more, expanded, and better programs of research training in environmental design research:
    a. Develop new graduate training programs that stress the holism and unique features of the entire field of environmental-behavior studies at all levels and scales of analysis and application.
    b. Encourage the development of more, expanded, and better doctoral research programs.
    c. Encourage the development of programs of environment-behavior interfacing with other disciplines (e.g., programs in environmental psychology, environmental sociology, social geography, natural resource management, social planning, landscape architecture, urban design, architecture, and interior design).
    d. Encourage graduate programs in the environmental professions (usually master's level programs) to have research thesis requirements that include the use of environment–behavior research information in environmental problem solving.
2. Encourage the development of new and expanded programs in continuing education for environmental design research:
    a. Develop new continuing education programs for post-occupancy evaluation, facility programming, applied research methods, and environment-behavior consulting for professionals wishing new or expanded training in environmental design research.
    b. Develop continuing education programs for professional decision makers (e.g., the urban design program in New York City).
    c. Investigate the feasibility of developing a continuing education program in conjunction with the annual organizational meetings in the field (as was done at the 1984 EDRA meetings).
    d. Develop internship programs focused on the use of research in environmental decision making.

3. Encourage the teaching of understanding and using research at the undergraduate level in professional schools and in the application of research to practical human-environmental problems in social and behavioral science departments.

INITIATING PROGRAMS WITHIN SCHOLARLY AND PROFESSIONAL ASSOCIATIONS

The sixth task facing the field in its efforts to implement more and better programs of research is suggested by the other five. It is to rethink the role, functions, and structures of the scholarly and professional associations that represent the field. All of the associations representing portions of the field suffer from two basic, and competing weaknesses. The first is that they attempt to do too many things without a clear sense of direction. The second is that they are unable to plan for long-range incremental actions while working with small budgets and the good will of their members. This sixth and final objective, therefore, is to look at the associations representing aspects of the field and to initiate new programs and processes with them that will better serve the research and applied needs of the field. This objective cannot be broached in any great depth in the present document. It calls for a study of its own (a good example is *Crisis in Architecture*, MacEwen, 1974, produced for the Royal Institute of British Architects). However, some beginning recommendations in strategic areas follow:

1. Establish a full-time Washington, D.C.-based organization with a full-time professional staff committed to the stimulation and development of the environmental design research field and to influencing policy, planning, design, and education. Study the feasibility of joining together with other similar sized organizations for the purpose of administrative efficiency and economies of scale (see AIA, 1982).
2. Join together in an interorganizational effort of environmental design and social science associations to encourage and develop interorganizational cooperation in the field of environmental design research (e.g., those in Appendix B). Develop an active commitment among these associations to engage in political action on behalf of the field, to study and know the timing and ways in which federal decisions are made, and to work effectively within this arena for the betterment of environmental design research and the quality of human life (e.g., analogous to or in concert with COSSA).
3. Organizations and associations in the field of environment-behavior research and applications are encouraged to initiate a series of organizational activities to benefit research and application in the field.

a. Establish data banks of research findings, information storage, retrieval, and dissemination.
b. Establish data banks listing names of researchers, consultants, and practitioners in the field who are doing environmental design research, translating it, and using it.
c. Expand organizational awards programs recognizing excellence in research and applications of environmental design research.
d. Expand publications to include annual reviews of progress and the publication of a series of research monographs for the dissemination of research information, findings, and recommendations to the environmental professionals charged with policy planning and design decision making.
4. The field of environment–behavior–design research should update this agenda on a regular five-year basis, using a panel of experts in the field prepared under the direction of a project director, and expand the agenda in intervening years through annual symposia, workshops and conferences.

NOTES

1. As Francis Ventre has pointed out on many occasions to the EDRA Board of Directors, one cannot just write an order to the world at large and expect results. These six general objectives are meant to suggest a context for particular objectives and strategies addressed to specific segments of opinion and influence.

2. This recommendation comes from the director of another association's Washington-based office, the Design and Environment Division of the Gerontological Society of America. Our thanks to the Director, Paul Taylor, for sharing his insights with the panel.

APPENDIXES

A:
ORGANIZATION AND PROCEDURE
OF THE PROJECT

This document was prepared by the authors for, and under the direction of, the EDRA Board of Directors. Support was given by an organizational Design Exploration/Research grant to EDRA from the Design Arts Program of the National Endowment for the Arts (Gary T. Moore and Sandra C. Howell, Project Directors). Conceptual directions were suggested by a Task Force or Research Agenda (1979-81 chaired by Sandra Howell, and including John Archea, Robert Gutman, William Ittelson, and Gary Moore), by four successive Boards of Directors (1979-80 chaired by John Archea, 1980-81 chaired by Gary Moore, 1981-82 chaired by Robert Gutman, and 1982-83 chaired by Jay Farbstein) and by a Research Agenda Review Committee (1981-82 chaired by Daniel Stokols and consisting of Stephen Margulis and Louis Sauer).

Throughout the process, consideration was given to ensuring representative input from the entire field of environmental design research. First, calls for input were published in the quarterly EDRA newsletter, *Design Research News*.

Second, position papers were contributed by members of the Task Force, and two Task Force retreats were held in Charleston, South Carolina (November 1979) and Ames, Iowa (April 1981).

Third, a survey was conducted in 1980 of leading researchers, scholars, research administrators, and practitioners both inside and outside the Association. Requests for input were sent to 440 people, of whom 88 (20%) submitted letter responses, many accompanied by position papers. The largest response was from academic researchers (42%) and the lowest response from professionals (9%).

Fourth, four membership workshops were held at the 1981 annual conference (EDRA 12, Ames, Iowa, April 1981). About 250 members of the association were present at one or more of these sessions, of whom 50 made comments that were tape-recorded for later analysis.

Fifth, input was invited from the Interorganizational Task Force on Human-Environment Research and Applications. The Interorganizational Task Force represented 24 scholarly and professional associations including sections concerned with environment-behavior research and environmental design applications, such as EDRA and the environment-behavior sections of the American Psychological Association, the American Sociological Association, the American Planning Association, and so on.

Sixth, the report has been informed by a review of earlier research agendas (for example, those sponsored by the National Science Foundation, National Endowment for the Arts, American Institute of Architects Research Corporation, Association of Collegiate Schools of Architecture, and Architectural Research Centers Consortium) each involving the ideas of 20 to 50 different scholars.

Seventh, the project directors were invited to participate in other research agenda symposia and have brought opinions and specific research issues from those symposia to the present endeavor.

A content analysis was conducted of all of these inputs (over 150 separate items representing the views of over 200 individuals). A tabulation of the results is presented in Appendix B, and the complete list of contributors is in Appendix C.

Four drafts of this document were produced between July 1981 and October 1982, and were submitted for review to the Board, the Task Force, and the Review Committee. In addition, a panel of advisors was asked to give a critical reading to the second draft. The panel consisted of Irwin Altman, Robert Bechtel, Michael Brill, John Eberhard, Robert Gutman, John Habraken, Gary Hack, Robert Marans, Stephen Margulis, Daniel Stokols, Francis Ventre, John Zeisel, and Ervin Zube.

Finally, drafts of this report have been reviewed and approved by the Task Force, the Review Committee, and the 1982 EDRA Board of Directors.

B:
TABULATION OF RESPONSES
TO THE EDRA SURVEY

Below is a sample tabulation of the respondents to the Environmental Design Research Association (EDRA) Survey of Research Needs and Directions and their responses.

TABLE 2: Written Responses to the EDRA Survey of Research Needs and Directions

	Surveys Mailed	Number of Responses	Average % Return
Academic researchers	135	57	42
Design professionals	165	15	9
Research administrators	140	16	11
Totals	440	88	20

TABLE 3: Oral Responses to the EDRA Survey*

	Number of Responses
EDRA 12 workshops	50
Interorganizational task force	9
Board of directors	7
Total	66

*Tape recorded from new people not previously counted as respondents to the letter survey.

TABLE 4: Reports Submitted as Part of the EDRA Survey

	Reports	Number of People
NEA Belmont Retreat Reports	11	12
NSF Reports[a]	8	51
Other Reports[b]	5	60
Totals	24	123

[a]One report represented 43 individuals.
[b]One report represented 56 individuals.

TABLE 5: Total Available Items in the EDRA Survey

	Responses	Number of People
Written responses	88	88
Oral responses	66	66
Reports	24	123
Totals	178	277

TABLE 6: Number of Respondents and Percentage of Responses Indicating Different Priority Areas and Issues

Area	Number of Respondents*	Percent of Responses	Issue	Percent of Responses
Theoretical issues	20	10	Development of theory	6
			Structural issues	4
Place issues	35	23	Macro-scale environments	10
			Housing	4
			Building systems and materials	3
			Workplaces	2
			Other place issues	4
Human group issues	30	16	Elderly	4
			Handicapped	4
			Children	3
			Ethnic groups	2
			Ordinary people	2
			Other group issues	1
Social and behavioral issues	16	8	Health and safety	3
			Crime and security	2
			Environmental perception, cognition, and meaning	1
			Other socio-behavioral issues	2
Process issues	54	25	Research implementation	10
			Design methods	6
			Research methods	4
			Other process issues	5
Contextual issues	35	18	Energy and behavior	4
			Futures and change	3
			Political-economic structures	3
			Preservation and reuse	1
			Technological assessment	1
			Other contextual issues	6

*Based on responses from 154 respondents (written plus oral responses). The totals are more than 154 as many people raised more than one research issue.

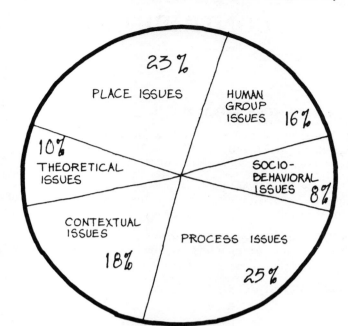

FIGURE 42 Illustrated percentage responses to the EDRA survey of research needs and priorities.

C: CONTRIBUTORS

PROJECT DIRECTORS AND AUTHORS

Gary T. Moore, Project Director and Senior Author
    Center for Architecture and Urban Planning Research,
    University of Wisconsin–Milwaukee

D. Paul Tuttle, Project Author
    Environment-Behavior Research Institute, University of Wisconsin–Milwaukee

Sandra C. Howell, Project Co-Director
    Department of Architecture, Massachusetts Institute of Technology

RESEARCH ASSISTANTS

Ellen M. Bruce
    Environment-Behavior Research Institute, University of Wisconsin–Milwaukee

Donna Duerk
    Department of Architecture, Massachusetts Institute of Technology

Patricia Gill
    Environment-Behavior Research Institute, University of Wisconsin–Milwaukee

EDRA TASK FORCE ON RESEARCH AGENDA, 1979–81

Sandra C. Howell, Chair
    Department of Architecture, Massachusetts Institute of Technology

John Archea
    Architectural Research Laboratory, Georgia Institute of Technology

Robert Gutman
    Department of Sociology, Rutgers University, and School of Architecture and Urban Planning, Princeton University

William H. Ittelson
    Environmental Psychology Program, University of Arizona

Gary T. Moore
    Center for Architecture and Urban Planning Research, University of Wisconsin–Milwaukee

142

EDRA RESEARCH AGENDA REVIEW COMMITTEE, 1981–82

Daniel Stokols, Chair
   Social Ecology Program, University of California, Irvine

Stephen T. Margulis
   Environmental Design Research Division, National Bureau of Standards, and Buffalo Organization for Social and Technological Innovation, Buffalo, New York

Louis Sauer
   Institute of Building Sciences, Carnegie-Mellon University

RESEARCH AGENDA ADVISORY PANEL, 1981–82

Ernest R. Alexander
   Department of Urban Planning, University of Wisconsin–Milwaukee

Irwin Altman
   College of Social and Behavioral Science, University of Utah

John Archea
   Architectural Research Laboratory, Georgia Institute of Technology

Robert Bechtel
   Department of Psychology, University of Arizona, and Environmental Research and Development Foundation, Tucson, Arizona

John K. Boal
   Midwest Regional Office, U.S. Department of Housing and Urban Development

Michael Brill
   Buffalo Organization for Social and Technological Innovation, Buffalo, New York

John P. Eberhard
   Advisory Board on the Built Environment, National Academy of Sciences

Robert Gutman
   Department of Sociology, Rutgers University, and School of Architecture, Princeton University

Gary Hack
   Department of Urban Studies and Planning, Massachusetts Institute of Technology

N. John Habraken
    School of Architecture and Planning, Massachusetts Institute of Technology

Paul Laseau
    College of Architecture and Planning, Ball State University

Robert W. Marans
    Institute for Social Research, University of Michigan, and College of Architecture and Planning, University of Michigan

Stephen T. Margulis
    Environmental Design Research Division, National Bureau of Standards, and Buffalo Organization for Social and Technological Innovation, Buffalo, New York

William J. Mitchell
    Urban Design Program, University of California, Los Angeles

Arthur H. Patterson
    Man–Environment Relations Program, Pennsylvania State University

Leanne G. Rivlin
    Environmental Psychology Program, City University of New York

Robert Shibley
    Solar Energy Program, U.S. Department of Energy

Daniel Stokols
    Program in Social Ecology, University of California, Irvine

Francis T. Ventre
    Division of Environmental Design, National Bureau of Standards, and Division of Civil and Environmental Engineering, National Science Foundation

John Zeisel
    Building Diagnostics Research, Cambridge, Massachusetts

Ervin H. Zube
    School of Renewable Natural Resources, University of Arizona

EDRA BOARDS OF DIRECTORS 1979–83

John Archea, Chair, 1979–80 (1977–81)
    Architectural Research Laboratory, Georgia Institute of Technology

Gary T. Moore, Chair, 1980–81 (1978–81)
    Environment-Behavior Research Institute, University of Wisconsin–Milwaukee

Robert Gutman, Chair, 1981–82 (1979–82)
Department of Sociology, Rutgers University, and School of Architecture and Urban Planning, Princeton University

Jay Farbstein, Chair, 1982–83 (1981–84)
Jay Farbstein and Associates, San Luis Obispo, California

Robin C. Moore, Chair, 1983–84 (1982–85)
Department of Landscape Architecture, North Carolina State University

Michael Brill (1977–80, 1982–85)
Buffalo Organization for Social and Technological Innovation, Buffalo, New York

Sandra C. Howell (1977–80)
Department of Architecture, Massachusetts Institute of Technology

William H. Ittelson (1978–81)
Environmental Psychology Program, University of Arizona

Min Kantrowitz (1980–83)
Min Kantrowitz and Associates, Albuquerque, New Mexico

Stephen T. Margulis (1979–82)
Environmental Design Research Division, National Bureau of Standards

Janet Reizenstein (1981–84)
Office of Planning, Development, and Research, University of Michigan Hospital

Louis Sauer (1980–83)
Institute of Building Sciences, Carnegie-Mellon University

Andrew P. Seidel (1979–82)
Department of Urban Planning, University of Texas at Arlington

Daniel Stokols (1980–83)
Program in Social Ecology, University of California, Irvine

Françoise Szigeti (1981–84)
The Environment Analysis Group, Ottawa (Canada)

Sue Weidemann (1977–80)
Housing Research and Development Program, University of Illinois

Ervin H. Zube (1982–85)
School of Renewable Natural Resources, University of Arizona

Willo P. White, Executive Officer
Environmental Design Research Association

RESPONDENTS TO THE EDRA LETTER SURVEY

The following is the complete list of respondents to the EDRA letter survey. The survey asked leading researchers, professionals, and research administrators to comment on three issues: (1) the philosophy of environmental design research; (2) priority research areas, topics, and issues; and (3) strategies for implementing the agenda and developing environmental design research capabilities over the next decade. Following is a list of the 88 people who submitted written responses to the survey and contributed the substance forming the basis of this document.

A

Cora Beth Abel
Adaptive Environments Center, Massachusetts College of Art

Douglas Amedeo
Department of Geography, University of Nebraska

John Archea
Architectural Research Laboratory, Georgia Institute of Technology

James R. Anderson
Housing Research and Development Program, University of Illinois

B

Michael Bakoy
Architecture/Research/Construction, Inc., Cincinnati, Ohio

Michael B. Barker
American Institute of Architects, Washington, D.C.

Robert Bechtel
Environmental Development and Research Foundation, Tucson, Arizona

Franklin D. Becker
Department of Design and Environmental Analysis, Cornell University

Robert M. Beckley
Department of Architecture, University of Wisconsin–Milwaukee

Lynn S. Beedle
Fritz Engineering Laboratory, Lehigh University

John K. Boal
> Midwest Regional Office, U.S. Department of Housing and Urban Development

Walter Bogan
> Office of Environmental Education, U.S. Department of Education

Sydney Brower
> School of Social Work and Community Planning, University of Maryland, and Department of City Planning, City of Baltimore

Michael Brill
> Buffalo Organization for Social and Technological Innovation, Buffalo, New York

C

Anthony James Catanese
> School of Architecture and Urban Planning, University of Wisconsin–Milwaukee

Richard E. Chenoweth
> Department of Landscape Architecture, University of Wisconsin–Madison

Uriel Cohen
> Department of Architecture, University of Wisconsin–Milwaukee

D

Louis J. D'Amore
> L.J. D'Amore & Associates, Architects, Montreal, Quebec (Canada)

Scott Danford
> Department of Environment Design, State University of New York at Buffalo

Gerald Davis
> The Environmental Analysis Group, Ottawa (Canada)

David Dibner
> Public Buildings Service, U.S. General Services Administration

F

Anthony J. Filipovitch
> Department of Psychology, Montana State University

Guido Francescato
    Department of Housing and Applied Design, University of Maryland

G

Barrie B. Greenbie
    Department of Landscape Architecture and Regional Planning, University of Massachusetts

James B. Griffin
    College of Architecture, University of Nebraska–Lincoln

Linda P. Groat
    Department of Architecture, University of Wisconsin–Milwaukee

H

N. John Habraken
    School of Architecture and Planning, Massachusetts Institute of Technology

John C. Haro
    Alfred Kahn Associates, Architects and Planners, Detroit, Michigan

Robert A. Hart
    Environmental Psychology Program, City University of New York

Samuel J. Hodges, III
    Community Conservation Research Division, U.S. Department of Housing and Urban Development

J

Abraham M. Jeger
    Department of Psychology, New York Institute of Technology

Joseph B. Juhasz
    Department of Environmental Design, University of Colorado

K

Robert W. Kates
    Graduate School of Geography, Clark University

L

Jon T. Lang
    Urban Design Program, University of Pennsylvania

Mary Helen Lorenz
Skidmore, Owings & Merrill, Architects/Planners, Boston, Massachusetts

Susan Low
Department of Anthropology, University of Pennsylvania

M

Suzanne R. McBride
Department of Educational Studies, University of Delaware

James McMenamin
The Physical Planning Group, Sabre Foundation, Santa Barbara, California

William Michelson
Program in Social Ecology, University of California, Irvine

William J. Mitchell
Architecture/Urban Design Program, University of California, Los Angeles

Mit Mitropoulos
Center for Advanced Studies in the Visual Arts, Massachusetts Institute of Technology

Robin C. Moore
People–Environment Relations, Consultants, San Francisco, California

D. Michael Murtha
D. Michael Murtha and Associates, Consultants, Bethesda, Maryland

Sherrill R. Myers
Department of Architecture, University of Wisconsin–Milwaukee

N

Lucille Nahemow
Division of Gerontology, New York University Medical Center

O

Jeffrey Oertel
Thorsen & Thorsen Associates, Architects/Planners, Minneapolis, Minnesota

Lance A. Olsen
Division of Human Services, College of Great Falls, Montana

P

H. McIlvaine Parsons
Human Resources Research Organization, Consultants, Alexandria, Virginia

Arthur H. Patterson
Man–Environment Relations Program, Pennsylvania State University

Beverly Field Pierz
Pierz Associates, Architecture and Interior/Environmental Design, Wethersfield, Connecticut

Joseph F. Pierz
Pierz Associates, Architecture and Interior/Environmental Design, Wethersfield, Connecticut

Michael John Pittas
Design Arts Program, National Endowment for the Arts

Peter Pollock
Solar Energy Research Institute, Golden, Colorado

Wolfgang F.E. Preiser
Institute for Environmental Education, University of New Mexico

R

Amos Rapoport
Department of Architecture, University of Wisconsin–Milwaukee

Victor Regnier
Housing Research and Development Program, University of Illinois, Urbana

Leanne Rivlin
Environmental Psychology Program, City University of New York

William Rock, Jr.
Department of Landscape Architecture, University of Toronto (Canada)

Richard Rush
Progressive Architecture, Stamford, Connecticut

S

Thomas F. Saarinen
> Department of Geography and Regional Development, University of Arizona

John Salman
> National Center for a Barrier-Free Environment, Washington, D.C.

Henry Sanoff
> School of Design, North Carolina State University

Roger Schluntz
> Association of Collegiate Schools of Architecture, Washington, D.C.

Donald E. Schmidt
> Department of Psychology, Virginia Polytechnic Institute and State University

David Seamon
> Department of Geography, University of Oklahoma

Robert Shibley
> Passive Solar Energy Division, U.S. Department of Energy

William Simms
> Department of City and Regional Planning, Ohio State University

M. Brewster Smith
> College of Social and Behavioral Sciences, University of California, Santa Cruz

Edward Steinfeld
> Department of Architecture, State University of New York at Buffalo

Daniel Stokols
> Program in Social Ecology, University of California, Irvine

Peter Suedfeld
> Department of Psychology, University of British Columbia (Canada)

François Szigeti
> The Environmental Analysis Group, Ottawa (Canada)

T

Ralph B. Taylor
> Center for Metropolitan Planning and Research, Johns Hopkins University

Paul S. Taylor
Gerontology Society of America, Washington, D.C.

Alan Temko
San Francisco Chronicle and Examiner, San Francisco, California

Philip Thiel
Department of Architecture, University of Washington

Richard Titus
Community Crime Prevention Division, U.S. Department of Justice

Jacqueline Tyrwhitt
Ekistics, Athens (Greece)

V

Harry Van Oudenallen
Department of Architecture, University of Wisconsin–Milwaukee

D.J. Van Rest
Interdisciplinary Higher Degrees Scheme Office, University of Aston, Birmingham (England)

Wayne Veneklasen
Civil Engineering Research Laboratory, U.S. Department of the Army, Champaign, Illinois

Francis T. Ventre
Environmental Design Research Division, U.S. National Bureau of Standards

W

John W. Wade
College of Architecture and Urban Studies, Virginia Polytechnic Institute and State University

Donald P. Watson
Department of Architecture, Yale University

Willo P. White
Environmental Design Research Association

James A. Wise
Departments of Architecture and Psychology, University of Washington

Lawrence P. Witzling
  School of Architecture and Urban Planning, University of Wisconsin–Milwaukee

Z

Ervin H. Zube
  School of Renewable Natural Resources, University of Arizona

WORKSHOP PARTICIPANTS, EDRA 12, AMES, IOWA, APRIL 1981

The following are the other EDRA members who contributed verbal ideas during one or more of the EDRA 12 workshops but are not on the above list of letter respondents.

Irwin Altman
  College of Social and Behavioral Sciences, University of Utah

Corwin Bennett
  Department of Industrial Engineering, Kansas State University

Michael I. Brooks
  College of Architecture and Planning, Iowa State University

F. Duncan Case
  Department of Textiles, Merchandising, and Design, University of Tennessee

Dana Cuff
  Department of Architecture, University of California, Berkeley

Jay Farbstein
  Jay Farbstein and Associates, Architects, San Luis Obispo, California

James B. Griffin
  College of Architecture, University of Nebraska

Min Kantrowitz
  Min Kantrowitz and Associates, Consultants, Albuquerque, New Mexico

Walter Kleeman
  Metagraphics Inc., Denver, Colorado

M. Powell Lawton
  Philadelphia Geriatric Center, Philadelphia, Pennsylvania

Stephen P. Margulis
    Environmental Design Research Division, U.S. National Bureau of Standards

Leon Pastalan
    Institute of Gerontology, University of Michigan

Janet Reizenstein
    Office of Research, Planning, and Development, University of Michigan Hospital

Andrew P. Seidel
    Department of Urban Planning, University of Texas at Arlington

Carol Weinstein
    Department of Educational Psychology, Rutgers University

Craig Zimring
    Architectural Research Laboratory, Georgia Institute of Technology

INTERORGANIZATIONAL TASK FORCE ON HUMAN-ENVIRONMENT RESEARCH AND APPLICATIONS

The following people attended a special half-day meeting of an Interorganizational Task Force on Human-Environment Research and Applications and offered their input to the report. Each person represented a scholarly or professional association concerned with aspects of environmental design research.

Gary T. Moore, Task Force Chair
    Environmental Design Research Association
    Environment-Behavior Research Institute, University of Wisconsin–Milwaukee

Irwin Altman
    American Psychological Association
    College of Social and Behavioral Sciences, University of Utah

Corwin Bennett
    Human Factors Society
    Department of Industrial Engineering, Kansas State University

Michael I. Brooks
    American Planning Association
    College of Design and Planning, Iowa State University

F. Duncan Case
    American Association of Housing Educators
    Department of Textiles, Merchandising, and Design, University of Tennessee

Kenneth H. Craik
    International Association of Applied Psychology
    Institute for Personality Assessment and Research, University of California, Berkeley

James B. Griffin
    Architectural Research Centers Consortium
    College of Architecture, University of Nebraska

Walter Kleeman
    American Society of Interior Design, and
    Institute of Building Designers, Metagraphics, Inc., Denver, Colorado

D. Michael Murtha
    Association for the Study of Man–Environment Relations
    D. Michael Murtha and Associates, Consultants, Bethesda, Maryland

Leon Pastalan
    Gerontological Society of America
    Institute for Gerontology, University of Michigan

Wolfgang F.E. Preiser
    Society for Human Ecology
    Institute for Environmental Education, University of New Mexico

Janet Reizenstein
    American Sociological Association
    Office of Planning, Research, and Development, University of Michigan Hospital

Thomas F. Saarinen
    Association of American Geographers
    Department of Geography and Regional Development, University of Arizona

Henry Sanoff
    Design Research Society (Great Britain)
    School of Design, North Carolina State University

Roger L. Schluntz
    Association of Collegiate Schools of Architecture
    Department of Architecture, University of Arizona

Martyn Symes
  International Association for the Study of People and Their Physical
    Surroundings
  Barlett School of Architecture, University of London (England)

Françoise Szigeti
  International Association for the Study of People and Their Physical
    Surroundings
  The Environmental Analysis Group, Consultants, Ottawa (Canada)

Ross Thorne
  Australian Association for People and the Man-Made Environment
  Architectural Psychology Research Unit, University of Sydney
    (Australia)

Ervin H. Zube
  American Society of Landscape Architects, and
  Council of Educators in Landscape Architecture
  School of Renewable Natural Resources, University of Arizona

OTHER AGENDA RECEIVED AND REVIEWED

In addition to individual contributors, we asked permission to incorporate ideas developed in other research agenda in environmental design, including those sponsored by the National Science Foundation, the National Endowment for the Arts, the American Institute of Architects Research Corporation, the Association of Collegiate Schools of Architecture, and the Architectural Research Centers Consortium. The agenda received and analyzed included the following:

The Belmont Retreat: Design research. *Progressive Architecture,* September 1979, *60*(9), 51–52.

Bender, R. & Rand, G. *Future Research Directions in Environmental Design.* Final Report to the National Science Foundation, College of Environmental Design, University of California, Berkeley, 1979.

Filipovitch, A.J. The child in the city: A research agenda. Paper presented at the Council of University Institutes of Urban Affairs, Omaha, Nebraska, March 1981.

Green, K. Researching the '80s. Special Issue of *Research and Design,* Spring 1980, *2*, (Whole No. 4).

Gutman, R. New strategies for the building community. Paper prepared for the Urban Design Educator's Symposium, San Juan, Puerto Rico, May 1981.

Jackson, J.B. Design research problems in the field of landscape studies. Paper prepared for the Belmont Design Research Retreat, Belmont, Maryland, July 1979.

Kilper, D. *National Science Foundation Agenda: Environmental Design Research.* Final Report to the National Science Foundation, College of Architecture, Virginia Polytechnic and State University, n.d.

King, J. & Guregian, S.A. *Environmental Design Research.* Final Report to the National Science Foundation, Architectural Research Laboratory, University of Michigan, 1979.

Kolata, G. NSF: Gains for applied research and engineering—What about architecture and design? *Design Matters* (National Endowment for the Arts).

Lagorio, H.J. *Environmental Research: NSF Program Options.* Washington, D.C.: National Science Foundation, 1976.

Martin, R.J.L. & Willeke, G.E. *The House That Jack Built: An Agenda for the Assessment of the Technologies of the Built Environment.* Final Report to the National Science Foundation, Architectural Research Laboratory, Georgia Institute of Technology, n.d.

Murtha, D.M. Program options for environmental design research. In S. Weidemann & J.R. Anderson (Eds.) *Priorities for Environmental Design Research.* Washington, D.C.: Environmental Design Research Association, 1978.

Pittas, M. Design exploration research and new NEA policy. Paper prepared for the Association of Collegiate Schools of Architecture Meetings, San Antonio, Texas, March 1980.

Progressive Architecture 29th Awards Program. Special Issue of *Progressive Architecture*, January 1982, *63* (Whole No. 1).

Rice Center for Community Design and Research. *Environmental Design Research.* Houston: Rice Center, 1976.

Ross, R.P. & Campbell, D.B. A review of the EDRA Proceedings: Where have we been? Where are we going? In W.E. Rogers & W.H. Ittelson (Eds.) *New Directions in Environmental Design Research.* Washington, D.C.: Environmental Design Research Association, 1979.

Salmen, J. *A Listing of Fifteen Suggested Priority Projects in the Area of Barrier Free Design.* Final Report to the Community Services Administration, National Center for a Barrier Free Environment, 1980.

Schluntz, R.L. *Survey on Sponsored Architectural Research* (2 Vols.). Washington, D.C.: Association of Collegiate Schools of Architecture, 1981.

Sims, B. Environmental design research: Some emerging directions. *Man-Environment Systems*, March 1978, *8*(2), 67–74.

Skidmore, Owings, & Merrill. *Program Options for Environmental Design Research*. Final Report to the National Science Foundation, Skidmore, Owings and Merrill, Boston, 1978.

Snyder, J.C. (Ed.) *ARCC-NSF Architectural Research Agenda*. Interim Report to the National Science Foundation, Architectural Research Centers, Consortium, Washington, D.C., 1981.

Sundstrom, E., Kastenbaum, D. & Konar-Goldband, E. *Physical Office Environments; Employee Satisfaction, and Job Performance*. Final Report to the AIA Research Corporation, Buffalo Organization for Social and Technological Innovation, Buffalo, N.Y., 1978.

Villecco, M. & Brill, M. *Environmental/Design Research: Concepts, Methods, and Values*. Washington, D.C.: National Endowment for the Arts, 1981.

Bibliography

Ajzen, I. & Fishbein, M. *Understanding Attitudes and Predicting Social Behavior.* Englewood Cliffs, N.J.: Prentice-Hall, 1980.

Alexander, C.W.J. *Notes on the Synthesis of Form.* Cambridge, Mass.: Harvard University Press, 1964.

Alexander, E.R. If planning isn't everything, maybe it's something. *Town Planning Review,* 1981, *52,* 14–25.

Allen, V.L. On convergent methodology: An aesthetic theory of school vandalism. In L. Van Ryzin (Ed.) *Research Methods in Behavior-Environment Studies.* Madison: University of Wisconsin-Madison, Institute for Environmental Studies, 1978.

Altman, I. (Chair) Theory of man-environment relations. In W.F.E. Preiser (Ed.) *Environmental Design Research,* Vol. 2. Stroudsburg, Penna: Dowden, Hutchinson & Ross, 1973.(a)

Altman, I. Some perspectives on the study of man-environment phenomena. *Representative Research in Social Psychology,* 1973, *4,* 109–186.(b)

Altman, I. *Environment and Social Behavior.* Monterey, Calif.: Brooks/Cole, 1975.

Altman, I. Environmental psychology and social psychology. *Personality and Social Psychology Bulletin,* 1976, *2,* 96–113.

Altman, I & Haythorne, W. Ecology of isolated groups. *Behavioral Science,* 1967, *12,* 169–182.

Altman, I, Rapoport, A., & Wohlwill, J.F. (Eds.) *Human Behavior and Environment,* Vol. 4. *Culture and the Environment.* New York: Plenum, 1980.

Altman, I. & Wohlwill, J.F. (Eds.) *Human Behavior and the Environment,* Vol. 3. *Children and the Environment.* New York: Plenum, 1978.

Altman, I. & Stokols, D. *Handbook of Environmental Psychology.* New York: Wiley, in press.

American Institute of Architects. *Direction '80s.* Washington, D.C.: American Institute of Architects, 1982.

American Institute of Architects Research Corporation. Researching the '80s. *Research and Design,* Spring 1980, *2,* 7–16.

American National Standards Institute. *Standards for Barrier-Free Design (ANSI A117.1).* Washington, D.C.: American National Standards Institute, 1979.

Anonymous. An interview with David Stea on 3'P's of environmental cognition: Perception, positivism, participation. *Middle Eastern Technical University Journal of the Faculty of Architecture,* 1980, *6,* 101–128.

Appleyard, D. Why buildings are known. *Environment and Behavior,* 1969, *1,* 131–156.

Appleyard, D. Styles and methods of structuring a city. *Environment and Behavior,* 1970, *2,* 100–117.

Appleyard, D. *Livable Streets.* Berkeley, Calif.: University of California Press, 1981.

Appleyard, D., Lynch, K. & Myer, J.R. *The View from the Road.* Cambridge, Mass.: MIT Press, 1964.

Archea, J. Establishing an interdisciplinary commitment. In B. Honikman (Ed.) *Responding to Social Change.* Stroudsburg, Penna.: Dowen, Hutchinson & Ross, 1975

Archea, J. Behavioral science impact to environmental policy: Applications to stairs. In P. Suedfild, J.A. Russell, L.M. Ward, F. Sziegeti, & G. Davis (Eds.) *The Behavioral Basis of Design,* Vol. 2, Stroudsburg, Penna.: Dowden, Hutchinson & Ross, 1977.

Archea, J. & Margulis, S.T. Environmental research inputs to policy and design programs: The case of preparation for involuntary relocation of the institutionalized elderly. In T.O. Byerts, S.C. Howell, & L.A. Pastalan (Eds.) *The Environmental Context of Aging: Life Styles, Environmental Quality, and Living Arrangements.* New York: Garland, 1979.

Bacon, E. *Design of Cities.* New York: Viking, 1967.

Baltes, P.B., Reese, H.W., & Nesselroade, J.R. *Life-Span Developmental Psychology: Introduction to Research Methods.* Monterey, Calif.: Brooks/Cole, 1977.

Barker, R.G. On the nature of the environment. *Journal of Social Issues,* 1963, *19,* 17–38.

Barker, R.G. Explorations in ecological psychology. *American Psychologist,* 1965, *20,* 1–14.

Barker, R.G. *Ecological Psychology: Concepts and Methods for Studying the Environment of Human Behavior.* Stanford, Calif.: Stanford University Press, 1968.

Barker, R.G. & Associates. *Habitats, Environments, and Human Behavior.* San Francisco: Jossey-Bass, 1978.

Barker, R.G. & Wright, H.F. *Midwest and its Children: The Psychological Ecology of an American Town.* New York: Row, Peterson, 1955.

Barnett, J. *Urban Design as Public Policy: Practical Methods for Improving Cities.* New York: Architectural Record Books, 1974.

Beal, F.H. Defining developmental objectives. In W.I. Goodman & E.C. Freund (Eds.) *Principles and Practice of Urban Planning.* Washington, D.C.: International City Managers Association, 1968.

Bechtel, R.B. *Enclosing Behavior.* Stroudsburg, Penna.: Dowden, Hutchinson & Ross, 1977.

Bechtel, R.B. What are post-occupancy evaluations?: A laymen's guide to the POE for housing. Unpublished draft final report to the U.S. Department of Housing and Urban Development, Environmental Research and Development Foundation, Tucson, n.d.

Bechtel R.B. *Introduction to Environment and Behavior Research.* Mss in preparation.

Becker, F.D. *Workspace: Creating Environments in Organizations.* New York: Praeger, 1981.

Beedle, L.S. Structures and the built environment: Societal needs. Paper prepared for the Airlie House Workshop to Define Needed Civil Engineering Research, Washington, D.C., June 1979.

Bell, D. *Coming of the Post-Industrial Society: A Venture in Social Forecasting.* New York: Basic Books, 1973.

The Belmont Retreat: Design research. *Progessive Architecture,* September 1979, *60*, 51–52.

Bender, R. & Rand G. *Future Research Directions in Environmental Design.* Final Report to the National Science Foundation, College of Environmental Design, University of California, Berkeley, 1979.

Bennett, C. *Spaces for People: Human Factors In Design.* Englewood Cliffs, N.J.: Prentice-Hall, 1977.

Bonta, J.P. *Architecture and its Interpretation.* New York: Rizzoli, 1979.

Bourestom, N. & Pastalan, L. *Forced Relocation: Setting, Staff, and Patient Effects.* Final report to the National Institute of Mental Health, April 1975.

Bourestom, N. & Tars, S. Alterations in life patterns following nursing home relocation. *Gerontologist,* 1974, *14*, 506–509.

Boyd, R.A. The luminous environment and its effects on man. In C.T. Larson (Ed.) *School Environments Research: Environmental Evaluations.* Ann Arbor, Mich.: University of Michigan, Architectural Research Laboratory, 1965.

Brabrook, A.T. The making of a department. *Urban Affairs Quarterly,* 1971, *6*, 277–296.

Brewer, M.B. & Collins, B.E. (Eds.) *Scientific Inquiry and the Social Sciences.* San Francisco: Jossey-Bass, 1981.

Brill, M. *Do Buildings Really Matter? Economic and Other Effects of Designing Behaviorally Supportive Buildings.* New York, N.Y.: Educational Facilities Laboratories, Academy for Educational Development, 1982.

Broadbent, G., Brent, R. & Jencks, C. (Eds.) *Signs, Symbols, and Architecture.* Chichester, England: Wiley, 1980.

Broadbent, G. & Ward, A. (Eds.) *Design Methods in Architecture.* London: Lund Humphries; and New York: Wittenborn, 1969.

Bronfenbrenner, U. The ecology of human development in retrospect and prospect. In H. McGurk (Ed.) *Ecological Factors in Human Development.* New York: North Holland, 1977.

Bunting, T.E. & Semple, T. McL. The development of an environmental response inventory for children. In A.D. Seidel & S. Danford (Eds.) *Environmental Design: Research, Theory and Application.* Washington, D.C.: Environmental Design Research Association, 1979.

Burden, E. *Architectural Delineation.* New York: McGraw-Hill, 1981.

Buttimer, A. Social space and the planning of residential areas. *Environment and Behavior,* 1972, *4*, 279–318.

Buttimer, A. *Values in Geography.* Washington, D.C.: Association of American Geographers, 1974.

Buttimer, A. & Seamon, D. (Eds.) *The Human Experience of Space and Place.* London: Croom/Helm, 1980.

Byerts, T.O. Reflecting user requirements in designing city parks. In M.P. Lawton, R.J. Newcomer & T.O. Byerts (Eds.) *Community Planning for an Aging Society.* Stroudsburg, Penna.: Dowden, Hutchinson & Ross, 1976.

Byerts, T.O., Howell, S.C., & Pastalan, L. (Eds) *The Environmental Context of Aging: Life Styles, Environmental Quality, and Living Arrangements.* New York: Garland, 1979.

Campbell, A., Converse, P.E. & Rodgers, W.L. *The Quality of American Life.* New York: Russell Sage Foundation, 1976.

Campbell, D.T. & Stanley, J.C. *Experimental and Quasi-Experimental Designs for Research.* Chicago: Rand-McNally, 1963.

Canter, D. *The Psychology of Place.* New York: St. Martin's Press, 1977.

Canter, D. & Canter, S. (Eds.) *Designing for Therapeutic Environments: A Review of Research.* New York: Wiley, 1979.

Caplan, N. A minimal set of conditions necessary for the utilization of social science knowledge in policy formulation at the national level. In C.H. Weiss (Ed.) *Using Social Research in Public Policy Making.* Lexington, Mass.: Lexington, 1977.

Caplan, N., Morrison, A. & Stembaugh, R. *The Use of Social Science Knowledge in Policy Decisions at the National Level.* Ann Arbor, Mich: University of Michigan, Institute for Social Research, 1975.

Carp, F.M. *A Future for the Aged.* Austin, Texas: University of Texas Press, 1966.

Carpman, J.R., Grant, M.A. & Simmons, D.A. Wayfinding in the hospital environment. *Journal of Environmental Systems,* 1983–84, *13,* 353–364.

Carson, D.H. The interactions of man and his environment. In C.T. Larson (Ed.) *School Environments Research,* Vol. 2. *Environmental Evaluations.* Ann Arbor, Mich.: University of Michigan, College of Architecture and Design, 1965.

Chapin, F.S., Jr. *Urban Land Use Planning.* Urbana, Ill.: University of Illinois Press, 1970.

Chase, R.A. (Chair) Theoretical issues in man-environment relations. In W.F.E. Preiser (Ed.) *Environmental Design Research,* Vol. 1. Stroudsburg, Penna.: Dowden, Hutchinson & Ross, 1973.

Clay, G. *Close Up: How to Read the American City.* New York: Praeger, 1973.

Clipson, C. & Werher, J. *Planning for Cardiac Care: A Guide for Planning and Design of Cardiac Care Facilities.* Ann Arbor, Mich.: Health Administration Press, 1974.

Cohen, S., Glass, D.C., & Singer, J.E. Apartment noise, auditory discrimination, and reading ability in children. *Journal of Experimental Social Psychology,* 1973, *9,* 407–422.

Cohen, U., Hetzer, J., Janiuk, R. & Tascioglu, G. Information needs and information use in architectural offices. *Wisconsin Architect,* October 1976, 1–4.

Cohen, U. & Moore, G.T. The organization and communication of behaviorally-based research information. In L. van Ryzin (Ed.) *Behavior-Environment Research Methods.* Madison: University of Wisconsin-Madison, Institute for Environmental Studies, 1977.

Conway, D. *Social Science and Design: A Process Model for Architect and Social Scientist Collaboration.* Washington, D.C.: American Institute of Architects, 1973.

Cook, T.D. & Campbell, D.T. The design and conduct of quasi-experiments and true experiments in field settings. In M.D. Dunnette (Ed.) *Handbook of Industrial and Organizational Psychology.* Chicago: Rand-McNally, 1976.

Cook, T.D. & Campbell, D.T. *Quasi-Experimentation: Design and Analysis Issues in Field Settings.* Chicago: Rand-McNally, 1979.

Cooper, C.C. Resident dissatisfaction in multi-family housing. In W.M. Smith (Ed.) *Behavior, Design, and Policy Aspects of Human Habitats.* Green Bay, Wisc.: University of Wisconsin-Green Bay Press, 1972.

Cooper, C.C. *Easter Hill Village: Some Social Implications of Design.* New York: Free Press, 1975.

Craik, K.H. The comprehension of the everyday physical environment. *Journal of the American Institute of Planners,* 1968, *34,* 27–37.

Craik, K.H. (Chair) Environmental dispositions and preferences. In J. Archea & C.M. Eastman (Eds.) *EDRA Two: Proceedings of 2nd Annual Environmental Design Research Association Conference.* Stroudsburg, Penna.: Dowden, Hutchinson & Ross, 1970. (a)

Craik, K.H. Environmental psychology. In K.H. Craik, B. Klunmurtz, R. Rosnow, R. Rosenthal, J.A. Cheyne & R.H. Walters (Eds.) *New Directions in Psychology,* Vol. 4. New York: Holt, Rinehart and Winston, 1970. (b)

Craik, K.H. Individual variations in landscape description. In E.H. Zube, R.O. Brush & J.C. Fabos (Eds.) *Landscape Assessment: Values, Perceptions and Resources.* Stroudsburg, Penna.: Dowden, Hutchinson & Ross, 1975.

Craik, K.H. The personality research paradigm in environmental psychology. In S. Wapner, S.B. Cohen & B. Kaplan (Eds.) *Experiencing the Environment.* New York: Plenum, 1976.

Craik, K.H. Multiple scientific paradigms in environmental psychology *International Journal of Psychology,* 1977, *12.*

Craik, K.H. & Zube, E.H. *Issues in Perceived Environmental Quality Research.* Amherst, Mass.: University of Massachusetts, Institute for Man and Environment, 1975.

Craik, K.H. & Zube, E.H. (Eds.) *Perceiving Environmental Quality: Research and Applications.* New York: Plenum, 1976.

Cuff, D. Negotiating architecture. In A.E. Osterberg, C.P. Tiernan & R.A. Findlay (Eds.) *Design Research Interactions.* Washington, D.C.: Environmental Design Research Associations, 1981.

Cuff, D. The context for design: Six characteristics. In P. Bart, A. Chen & G. Francescato (Eds.) *Knowledge for Design.* Washington, D.C.: Environmental Design Research Association, 1982.

Cuff, D. & Wittman, F. Studying ourselves: Research on the design professions—implications for education and practice. In P. Bart, A. Chen & G. Francescato (Eds.) *Knowledge for Design.* Washington, D.C.: Environmental Design Research Association, 1982.

Daish, J., Gray, J., Kernohan, D. & Salmond, A. *Post-Occupancy Evaluation: Trial Study 3.* Wellington, New Zealand: Victoria University of Wellington, School of Architecture, 1981.

D'Amore, L.J. & Associates. *Survey on Sponsored Architectural Research: Analysis of Activities, Directions, and Issues.* Ottawa: Environment Canada, 1979.

Danford. S. Subjective responses in environmental design research: The controversy in perspective. In P. Bart, A. Chen & G. Francescato (Eds.) *Knowledge for Design.* Washington, D.C.: Environmental Design Research Association, 1982.

Design Research News. Washington, D.C.: Environmental Design Research Association, 1969 - present.

Dixon, J.M. (Ed.) 29th Annual Awards Issue. Special issue of *Progressive Architecture,* January 1982, *63* (Whole No. 1).

Downs, A. *Inside Bureaucracy.* Boston: Little Brown, 1967.

Downs, A. *Urban Problems and Prospects.* Chicago: Rand McNally, 1970.

Downs, R.M. Personal constructions of personal construct theory. In G.T. Moore & R.G. Golledge (Eds.) *Environmental Knowing: Theories, Research, and Methods.* Stroudsburg, Penna.: Dowden, Hutchinson & Ross, 1976.

Downs. R.M. & Stea, D. (Eds.) *Image and Environment: Cognitive Mapping and Spatial Behavior.* Chicago: Aldine, 1973.

Doxiadis, C.A. *Anthropopolis: City for Human Development.* New York: Norton, 1974.

Dreyfuss, H. *The Measure of Man: Human Factors in Design.* New York: Whitney Library of Design, 1966.

Duffy, F., Cave, C., & Worthington, J. (Eds.) *Planning Office Space.* London: Architectural Press; and New York: Nichols, 1976.

Duhl, L.J. (Ed.) *The Urban Condition.* New York: Clarion, 1963.

Eastman, C.M. (Ed.) *Spatial Synthesis in Computer-Aided Building Design.* New York: Halsted, 1975.

Elsner, G.H. & Smardon, R.C. *Our National Landscape.* Berkeley, Calif.: U.S. Department of Agriculture, Forest Service, Pacific South West Forest and Range Experiment Station, 1979.

Environment and Behavior. Beverly Hills, Calif.: Sage Publications, 1969 - present.

Environmental Design Research Association Proceedings (various titles and editors). Washington, D.C.: Environmental Design Research Association, 1969-present.

Epp, G., Georgapulos, D. & Howell, S. Monitoring environment behavior research. *Journal of Architectural Research,* 1979, *7,* 2–21.

Evans, G.W. Environmental cognition. *Psychological Bulletin,* 1980, *88,* 259–287.

Evans, G.W. & Howard, R.B. Personal space. *Psychological Bulletin,* 1973, *80,* 334–344.

Farbstein, J. & Wener, R.E. *Evaluation of Correctional Environments.* San Luis Obispo, Calif.: Farbstein/Williams & Associates, 1979.

Farbstein, J. & Wener, R.E. Evaluation of correctional environments. In G.T. Moore (Ed.) Applied Architectural Research: Post-Occupancy Evaluation of Buildings. Special issue of *Environment and Behavior,* 1982, *14,* 671–694.

Farr, L.E. Medical consequences of environmental home noises. In R. Gutman (Ed.) *People and Buildings.* New York: Basic Books, 1972.

Fava, S. Women's place in the new suburbia. In G.R. Wekerle, R. Peterson & D. Morley (Eds.) *New Space for Women.* Boulder, Colo., Westview, 1980.

Feldt, A.G., Marans, R.W. & Pastalan, L.A. Changing properties of retirement communities. In E. Smart (Ed.) *Housing for a Maturing Population.* Washington, D.C.: Urban Land Institute, 1983.

Festinger, L., Schachter, S. & Back, K.W. *Social Pressures in Informal Groups.* Stanford, Calif.: Stanford University Press, 1950.

Filipovitch, A.J. The child in the city: A research agenda. Paper presented at the Council of University Institutes of Urban Affairs, Omaha, Nebraska, March 1981.

Firey, W. Sentiment and symbolism as ecological variables. *American Sociological Review,* 1945, *10,* 140–148.

Fishbein, M. & Ajzen, I. *Belief, Attitude, Intention, and Behavior: An Introduction to Theory and Research.* Reading, Mass.: Addison-Wesley, 1975.

Foster, G.M. *Applied Anthropology.* Boston: Little Brown, 1969.

Fox, B.J., Fox, J. & Marans, R.W. Residential density and neighbor interaction. *Sociological Quarterly,* 1980, *21,* 349–359.

Francescato, G., Weidemann, S., Anderson, J.R. & Chenoweth, R. *Residents' Satisfaction in HUD-Assisted Housing: Design and Management Factors.* Washington, D.C.: U.S. Government Printing Office and U.S. Department of Housing and Urban Development, 1977.

Friedmann, A., Zimring, C. & Zube, E. (Eds.) *Environmental Design Evaluation.* New York: Plenum, 1978.

Gans, H.J. *The Urban Villagers.* New York: Free Press, 1959.

Gans, H.J. *The Levittowners.* New York: Random House, 1967.

Gans, H.J. *People and Plans.* New York: Basic Books, 1968.

Gans, H.J. *Popular Culture and High Culture: An Analysis and Evaluation of Taste.* New York: Basic Books, 1974.

Garling, T. Studies in visual perception of architectural spaces and rooms. *Scandinavian Journal of Psychology,* 1970, *11,* 133–145.

Gaskie, M.F. Toward workability of the workplace. *Architectural Record,* Mid-August 1980, *168,* 70–75.

Gero, J. (Ed.) *Computer Applications in Architecture.* London: Applied Science Publishers, 1977.

Glass, D.C. & Singer, J.E. *Urban Stress.* New York: Academic Press, 1972.

Goodman, W.I. & Freund, E.C. *Principles and Practice of Urban Planning.* Washington, D.C.: Internation City Managers' Association, 1968.

Green, K. Researching the '80s. Special issue of *Research and Design,* Spring 1980, *2,* (Whole No. 4).

Groat, L.P. Meaning in architecture: New directions and sources. *Journal of Environmental Psychology,* 1981, 73–83.

Groat, L.P. & Canter, D. Does post-modernism communicate? *Progressive Architecture,* December 1979, *60,* 84–87.

Gutman, R. New strategies for the building community. Paper presented at the Urban Design Educator's Symposium, San Juan, Puerto Rico, May 1981.

Hack, G. Research for urban design. In J.C. Snyder (Ed.) *Architectural Research.* New York: Van Nostrand Reinhold, 1984.

Hagino, G. & Ittelson, W.H. (Eds.) *Interaction Process Between Human Behavior and Environment.* Tokyo: Bunsei, 1980.

Hall, E.T. *The Silent Language.* Garden City, N.Y.: Doubleday/Anchor, 1959.

Hall, E.T. *The Hidden Dimension.* Garden City, N.Y.: Doubleday/Anchor, 1966.

Harris, L. & Associates. *The Steelcase National Study of Office Environments: Do They Work?* Grand Rapids, Mich.: Steelcase, 1978.

Hart, R.A. & Moore, G.T. The development of spatial cognition: A review. In R.M. Downs and D. Stea (Eds.) *Image and Environment: Cognitive Mapping and Spatial Behavior.* Chicago: Aldine, 1973.

Hayden, D. *Seven American Utopias.* Cambridge, Mass.: MIT Press, 1976.

Hayden, D. *The Grand Domestic Revolution: A History of Feminist Designs for American Homes, Neighborhoods, and Cities.* Cambridge, Mass.: MIT Press 1981.

Hayward, S.C. & Franklin, S.S. Perceived openness—enclosure of architectural spaces. *Environment and Behavior,* 1974, *6,* 37–52.

Helmreich, R. Evaluation of environments: Behavioral observations in an undersea habitat. In J. Lang, C. Burnette, W. Moleski & D. Vachon (Eds.) *Designing for Human Behavior.* Stroudsburg, Penna.: Dowden, Hutchinson & Ross, 1974.

Hershberger, R.G. A study of meaning in architecture. *Man and his Environment,* 1968, *1,* 6–7.

Hiatt, L.G. Reflections on an environmental psychologist's roles in design and architecture for older people. In A.D. Seidel & S. Danford (Eds.) *Environmental Design: Research, Theory and Application.* Washington, D.C.: Environmental Design Research Association, 1979.

Hill, A.B. *The Small Public Library: Design Guide, Site Selection, and Design Case Study.* Milwaukee: University of Wisconsin-Milwaukee, Center for Architecture and Urban Planning Research, 1980.

Honikman, B. *Responding to Social Change.* Stroudsburg, Penna.: Dowden, Hutchinson & Ross, 1975.

Howell, S.C. *Designing for Aging: Patterns of Use.* Cambridge, Mass.: MIT Press, 1980.

Hunt, M.E., Feldt, A.G., Marans, R.W., Pastalan, L.A., & Vakalo, K. *Retirement Communities: An American Original.* New York: Haworth, 1984.

Ittelson, W.H., Perception of the large-scale environment. *Transactions of the New York Academy of Sciences,* Series II, 1970, *32,* 807–815.

Ittelson, W.H., Proshansky, H.M. & Rivlin, L.G. Bedroom size and social interaction of the psychiatric ward. *Environment and Behavior,* 1970, *2,* 255–270.

Jackson, J.B. (Ed.) *Landscape,* Vol. 1–17, 1951 to 1968.

Jackson, J.B. *American Space.* New York: Norton, 1972.

Jackson, J.B. Design research problems in the field of landscape studies. Paper presented at the Belmont Design Research Retreat, Belmont, Maryland, July 1979.

Jacobi, M. & Stokols, D. The role of tradition in person-environment relations. Unpublished manuscript, Social Ecology Program, University of California, Irvine, 1981.

Jacobs, J. *The Death and Life of Great American Cities.* New York: Random House, 1961.

Jencks, C. *Modern Movements in Architecture.* Harmondsworth, England: Penguin, 1973.

Jencks, C. *The Language of Post-Modern Architecture* (Rev. ed.). New York: Rizzoli, 1981.

Kahn, H. & Bruce-Briggs, B. *Things to Come: Thinking about the Seventies and Eighties.* New York: MacMillan, 1972.

Kates, R.W. & Wohlwill, J.F. (Eds.) Man's Response to the Physical Environment. Special issue of the *Journal of Social Issues,* 1966, *22* (Whole No. 1).

Kemper, A. *Presentation Drawings by American Architects.* New York: Wiley, 1977.

Kilper, D. *National Science Foundation Agenda: Environmental Design Research.* Final Report to the National Science Foundation, College of Architecture, Virginia Polytechnic and State University, n.d.

King, J. & Guregian, S.A. *Environmental Design Research.* Final Report to the National Science Foundation, Architectural Research Laboratory, University of Michigan, 1979.

Kira, S. *The Bathroom: Criteria for Design.* New York: Viking, 1966.

Kleeman, W. *The Challenge of Interior Design.* Boston: CBI Publishing, 1981.

Knight, R.C. & Campbell, D.E. Environmental evaluation research: Evaluator roles and inherent social commitments. *Environment and Behavior,* 1980, *12,* 520–532.

Korobkin, B.J. *Images For Design: Communicating Social Science Research to Architects.* Cambridge, Mass.: Harvard University, Architectural Research Office, 1976.

Kuhn, T. *The Structure of Scientific Revolutions.* Chicago: University of Chicago Press, 1962.

Kuller, R. (Ed.) *Architectural Psychology.* Lund, Sweden: Studentlitteratur ab; and Stroudsburg, Penna.: Dowden, Hutchinson & Ross, 1973.

Lagorio, H.J. *Environmental Design Research: NSF Program Options.* Washington, D.C.: National Science Foundation, 1976.

Langdon, F.J. *Modern Offices: A User Survey.* London: Her Majesty's Stationary Office, 1966.

Lansing, J.B., Marans, R.W. & Zehner, R.B. *Planned Residential Environments.* Ann Arbor, Mich.: University of Michigan, Institute for Social Research, 1970.

Larson, C.T. (Ed.) *School Environments Research* (3 vols.). Ann Arbor, Mich.: University of Michigan, College of Architecture and Design, 1965.

Lawton, M.P. Competence, environmental press, and the adaptation of older people. In P.G. Windley, T.O. Byerts & F.G. Ernst (Eds.) *Theory Development in Environment and Aging.* Washington, D.C.: Gerontological Society, 1975.

Lawton, M.P. *Environment and Aging.* Monterey, Calif.: Brooks/Cole, 1980.

Lawton, M.P., Newcomer, R.J. & Byerts, T.O. (Eds.) *Community Planning for an Aging Society.* Stroudsburg, Penna.: Dowden, Hutchinson & Ross, 1977.

Lawton, M.P., Windley, P.G. & Byerts, T.O. (Eds.) *Aging and the Environment: Theoretical Approaches.* New York: Springer, 1982.

Lazarsfeld, P.F., Sewell, W. & Wilensky, H. (Eds.) *The Uses of Sociology.* New York: Basic Books, 1967.

Lazarus, R.S. & Launier, R. Stress related transactions between person and environment. In L.A. Pervin & M. Lewis (Eds.) *Perspectives in International Psychology.* New York: Plenum, 1978.

Levin, H. & Duhl, L.J. Architectural research and the impact of buildings on health. In J.C. Snyder (Ed.) *Architectural Research.* New York: Van Nostrand Reinhold, 1984.

Levy-Leboyer, C. *Psychology and Environment.* Paris: Presses Universitaires de France, 1979; and Beverly Hills, Calif.: Sage, 1982.

Lewin, K. *Principles of Topological Psychology.* New York: McGraw-Hill 1936.

Lewin, K. Behavior and development as a function of the total situation. In L. Carmichael (Ed.) *Manual of Child Psychology.* New York: Wiley, 1946.

Lewin, K. *Field Theory in the Social Sciences: Selected Theoretical Papers* (Ed. by D. Cartwright). New York: Harper & Row, 1951.

Lindblom, C.L. *The Policy-Making Process.* Englewood Cliffs, N.J.: Prentice-Hall, 1968.

Lowenthal, D. Geography, experience, and imagination: toward a geographical epistemology. *Annals of the Association of American Geographers,* 1961, *51*, 241–260.

Lunden, G. Environment problems of office workers. *Build International,* March-April 1972, *5*, 90–93.

Lynch, K. *The Image of the City.* Cambridge, Mass.: MIT Press, 1960.

Lynch, K. *What Time Is This Place?* Cambridge, Mass.: MIT Press, 1973.

Lynch, K. *Managing the Sense of a Region.* Cambridge, Mass.: MIT Press, 1976.

Lynch, K. *A Theory of Good City Form.* Cambridge, Mass.: MIT Press, 1981.

MacEwen, M. *Crisis in Architecture.* London: Royal Institute of British Architects Publications, 1974.

MacLane, S. Total reporting for scientific work. *Science,* 1980, *210*, 158–163.

Mahajan, B.M. & Beine, W.B. Impact attenuation performance of surfaces installed under playground equipment. Project paper prepared at the Center for Consumer Product Technology, U.S. National Bureau of Standards, September 1978.

Manning, P. *Office Design: A Study of Environment.* Liverpool, England: Pilkington Research Unit, 1965.

Marans, R.W. Perceived quality of residential environments: Some methodological issues. In K.H. Craik & E.H. Zube (Eds.) *Perceiving Environmental Quality: Research and Applications.* New York: Plenum, 1976.

Marans, R.W. *The Determinants of Neighborhood Quality: An Analysis of the 1976 Annual Housing Survey.* Washington, D.C.: U.S. Government Printing Office and U.S. Department of Housing and Urban Development, 1979.

Marans, R.W. Evaluation in architecture. In J.C. Snyder (Ed.) *Architectural Research.* New York: Van Nostrand Reinhold, 1984.

Marans, R.W. & Spreckelmeyer, K.F. *Evaluating Built Environments: A Behavioral Approach.* Ann Arbor, Michigan: University of Michigan, Institute for Social Research, 1981.

Marans, R.W. & Spreckelmeyer, K.F. Measuring overall architectural quality: a component of building evaluation. In G.T. Moore (Ed.) Applied Architectural Research: Post-Occupancy Evaluation of Buildings. Special issue of *Environment and Behavior,* 1982, *14*, 652–670.

Margulis, S.T. How environmental research may affect the technical provisions and enforcement of regulations. In P. Cooke (Ed.). *Research and Innovation in the Building Regulatory Process.* Washington, D.C.: National Bureau of Standards, NBS Special Publication 473, 1977.

Marshall, N.J. Dimensions of privacy preferences. *Multivariate Behavior Research,* 1974, *9,* 255–272.

Marshall, R. & Ruegg, R. *Efficient Allocation of Research Funds: Economic Evaluation Methods with Case Illustrations in Building Technology.* Washington, D.C.: U.S. Government Printing Office, 1980.

Martin, R.J.L. & Willeke, G.E. *The House that Jack Built: An Agenda for the Assessment of the Technologies of the Built Environment.* Final Report to the National Science Foundation, Architectural Research Laboratory, Georgia Institute of Technology, 1978.

Maslow, A.B. *Motivation and Personality.* New York: Harper & Row, 1954.

McKechnie, G.E. *Manual for the Environmental Response Inventory.* Palo Alto, Calif.: Consulting Psychologists Press, 1974.

Meadows, D.H., Meadows, D.L., Randers, J., & Behrens, W.W. *The Limits to Growth.* New York: Universe Books, 1972.

Meinig, D.W. (Ed.) *The Interpretation of Ordinary Landscapes.* New York: Oxford University Press, 1979.

Meltsner, A.J. *Policy Analysts in the Bureaucracy.* Berkeley, Calif.: University of California Press, 1976.

Mercer, C. *Living in Cities: Psychology and the Urban Environment.* Harmondsworth, England: Penguin, 1975.

Merrill, J.L., Jr. *Factors Influencing the Use of Behavioral Research in Design.* Unpublished Ph.D. dissertation, University of Michigan, 1976.

Merton, R.K. The position of sociological theory. *American Sociological Review,* 1948, *13,* 164–168.

Merton, R.K. Theories of the middle range. *Social Theory and Social Structure* (Rev. ed.). New York: Free Press, 1957.

Michelson, W. (Ed.) *Behavioral Research Methods in Environmental Design.* Stroudsburg, Penna.: Dowden, Hutchinson & Ross, 1975.

Michelson, W. *Man and His Urban Environment: A Sociological Approach.* Reading, Mass.: Addison-Wesley, 1976.

Michelson, W. From congruence to antecedent conditions: a search for the basis of environmental improvement. In D. Stokols (Ed.) *Perspectives on Environment and Behavior.* New York: Plenum, 1977a.

Michelson, W. *Environmental Choice, Human Behavior, and Residential Satisfaction.* New York: Oxford University Press, 1977b.

Milgram, S. The experience of living in cities. *Science,* 1970, *67,* 1461–1468.

Mischel, W. On the future of personality measurement. *American Psychologist,* 1977, *32,* 246–254.

Mitchell, W.J. *Computer-Aided Architectural Design.* New York: Petrocelli/ Charter, 1977.

Moore, G.T. (Chair) Symposium on conceptual issues in environmental cognition research. In W.J. Mitchell (Ed.) *Environmental Design: Research and Practice,* Vol. 2. Los Angeles: University of California, School of Architecture and Urban Planning, 1972.

Moore, G.T. Theory and research on the development of environmental knowing. In G.T. Moore & R.G. Golledge (Eds.) *Environmental Knowing: Theory, Research, and Methods.* Stroudsburg, Penna.: Dowden, Hutchinson & Ross, 1976.

Moore, G.T. Environment-behavior studies. In J.C. Snyder & A.J. Catanese (Eds.) *Introduction to Architecture.* New York: McGraw-Hill, 1979. (a)

Moore, G.T. Knowing about environmental knowing: the current state of theory and research on environmental cognition. *Environment and Behavior,* 1979, *11,* 33–70. (b)

Moore, G.T. Research, design, and evaluation for people-in-environments. In J.C. Snyder & A.J. Catanese (Eds.) *Introduction to Architecture.* New York: McGraw-Hill, 1979. (c)

Moore, G.T. Holism, environmentalism, and ecological validity: Comments on environmental transition and convergent methodology. *Man-Environment Systems,* 1980, *10,* 1–11. (a)

Moore, G.T. EDRA at the cross-roads in the 1980s. Foreword to A. Osterberg, C.P. Tiernan, & R.A. Findlay (Eds.) *Design Research Interactions.* Washington, D.C.: Environmental Design Research Association, 1980. (b)

Moore, G.T. A methodological critique of child-environment studies. Paper presented at the Environmental Design Research Association Conference. College Park, Maryland, April 1982. (a)

Moore, G.T. Standards on playground design and safety. Unpublished manuscript, Environment-Behavior Research Institute, University of Wisconsin-Milwaukee, 1982. (b)

Moore, G.T. Environment-behavior theory and research in architecture. In J.C. Snyder (Ed.) *Architectural Research.* New York: Van Nostrand Reinhold, 1984.

Moore, G.T. The state-of-the-art in play environment research and applications. In J. Frost (Ed.) *When Children Play*. Washington, D.C.: Association for Childhood Education International, in press. (a)

Moore, G.T. Research opportunities in environmental gerontology. In P. Heyer (Ed.) *Architecture and the Future*. Washington, D.C.: Association of Collegiate Schools of Architecture, in press. (b)

Moore, G.T., Cohen, U., & McGinty, T. *Planning and Design Guidelines: Child Care Centers and Outdoor Play Environments* (7 vols.). Milwaukee: University of Wisconsin-Milwaukee, Center for Architecture and Urban Planning Research, 1979 ad seriatum.

Moore, G.T., Cohen, U., Oertel, J., & Van Ryzin, L. *Designing Environments for Handicapped Children*. New York: Educational Facilities Laboratories, 1979.

Moore, G.T. & Golledge, R.G. (Eds.) *Environmental Knowing: Theories, Research, and Methods*. Stroudsburg, Penna.: Dowden, Hutchinson & Ross, 1976.

Moore, G.T. & Marans, R.W. The process of design and construction: building evaluation. In J.C. Snyder (Ed.) *An Agenda for Architectural Research 1982*. Washington, D.C.: Architectural Research Centers Consortium, n.d. (1982).

Moore, G.T., Tuttle, D.P., Howell, S.C., & Durek, D. Environmental design research for the 1980s: towards an agenda for research. In A. Osterberg, C.P. Tiernan & R.A. Findlay (Eds.) *Design Research Interactions*. Washington, D.C.: Environmental Design Research Association, 1980.

Moore, R.C. *Study Report: The Social Benefits and Assessment of Local Urban Open Space*. Washington, D.C.: American Institute of Architects Research Corporation, 1978.

Moos, R.H. Conceptualizations of human environments. *American Psychologist*, 1973, *28*, 652–665.

Moos, R. *The Human Context: Environmental Determinants of Behavior*. New York: Wiley, 1976.

Morris, E.W. & Winter, M. *Housing, Family, and Society*. New York: Wiley, 1978.

Moynihan, D.P. State vs. Academe. *Harpers*, 1980, *261*(1567), 31–40.

Muller, R.A. Innovation and scientific funding. *Science*, 1980, *209*, 880–883.

Murtha, D.M. Program options for environmental design research. In S. Weidemann & J.R. Anderson (Eds.) *Priorities for Environmental Design Research*. Washington, D.C.: Environmental Design Research Association, 1978.

Nahemow, L. & Lawton, M. Toward an ecological theory of adaptation and aging. In W.F.E. Preiser (Ed.) *Environmental Design Research*, Vol. 1. Stroudsburg, Penna.: Dowden, Hutchinson & Ross, 1973.

National Academy of Sciences. *Social and Behavioral Science Programs in the National Science Foundation*. Washington, D.C.: National Academy of Sciences, 1976.

National Bureau of Standards. *Time-Based Capabilities of Occupants to Escape Fires in Public Buildings.* Washington, D.C: U.S. Department of Commerce, National Bureau of Standards, 1982.

National Center for a Barrier Free Environment. *Research Report: A Listing of Fifteen Suggested Projects Which Should Have a High Priority for Research in the Area of Barrier Free Design.* Washington, D.C: National Center for a Barrier Free Environment, 1980.

National Fire Protection Association. *Life Safety Code.* Washington, D.C.: National Fire Protection Association, 1976.

National School Public Relation Association. *Violence and Vandalism.* Washington, D.C: National School Public Relations Association, 1975.

The NEA Belmont Retreat: design research. *Progressive Architecture,* September 1979, *60,* 51–52.

Newman, O. *Defensible Space: Crime Prevention Through Urban Design.* New York: Macmillan, 1972.

Newman, O. *Architectural Design for Crime Prevention.* Washington, D.C: U.S. Government Printing Office, 1973.

Newman, O. *Design Guidelines for Creating Defensible Space.* Washington, D.C: U.S. Government Printing Office, 1976.

Newman, O. *Community of Interest.* New York: Doubleday/Anchor, 1980.

Odum, H.T. *Environment, Power, and Society.* New York: Wiley-Interscience, 1971.

Osmond, H. Function as a basis of psychiatric ward design. *Mental Hospital,* 1957, *8,* 23–29.

Pablant, P. & Baxter, J.C. Environmental correlates of school vandalism. *Journal of the American Institute of Planners,* 1975, *41,* 270–279.

Palmer, M. *The Architect's Guide to Facility Programming.* Washington, D.C: American Institute of Architects; and New York: Architectural Record Books, 1981.

Parsons, H.McI. (Ed.) Environmental Design. Special issue of *Human Factors,* 1972, *14* (Whole No. 5).

Parsons, H.McI. Work environments. In I. Altman & J. Wohlwill (Eds.) *Human Behavior and Environment,* Vol. 1. New York: Plenum, 1976.

Patricios, N.N. An agentive model of person-environment relations. *International Journal of Environmental Studies,* 1978, *13,* 43–52.

Patterson, A.H. Methodological developments in environment-behavior research. In D. Stokols (Ed.) *Perspectives on Environment and Behavior: Theory, Research, and Applications.* New York: Plenum, 1977.

Patton, M. *Qualitative Evaluation Methods.* Beverly Hills, Calif.: Sage, 1980.

Peet, R. (Ed.) *Radical Geography: Alternative Viewpoints on Contemporary Social Issues.* Chicago: Maaroufa Press, 1977.

Piaget, J. *The Development of Thought: Equilibration of Cognitive Structures.* New York: Viking, 1975.

Pittas, M. Design exploration research and new NEA policy. Paper presented at the Association of Collegiate Schools of Architecture Meetings, San Antonio, Texas, March 1980.

Preiser, W.F.E. (Ed.) *Facility Programming.* Stroudsburg, Penna.: Dowden, Hutchinson & Ross, 1978.

President's Commission for a National Agenda for the Eighties. *The Quality of American Life in the Eighties.* Washington, D.C: U.S. Government Printing Office, 1980.

Proceedings of the Annual Environmental Design Research Association Conferences. Washington, D.C: Environmental Design Research Association, 1969-present.

Progressive Architecture 29th Awards Program. Special Issue of *Progressive Architecture,* January 1982, *63* (Whole No. 1).

Proshansky, H.M. Methodology in environmental psychology: Problems and issues. *Human Factors,* 1972, *14,* 451–460.

Proshansky, H.M. Comment on environmental psychology and social psychology. *Personality and Social Psychology Bulletin,* 1976, *2,* 356–363.

Proshansky, H.W. & Altman, I. Overview of the field. In W.P. White (Ed.) *Resources in Environment and Behavior.* Washington, D.C: American Psychological Association, 1979.

Proshansky, H.M. & O'Hanlon, T. Environmental psychology: Origins and development. In D. Stokols (Ed.) *Perspectives on Environment and Behavior.* New York: Plenum, 1977.

Rabinowitz, H.Z. Post-occupancy evaluation. In J.C. Snyder & A.J. Catanese (Eds.) *Introduction to Architecture.* New York: McGraw-Hill, 1979.

Rapoport, A. *House Form and Culture.* Englewood Cliffs, N.J.: Prentice-Hall, 1969.

Rapoport, A. An approach to the construction of man–environment theory. In W.F.E. Preiser (Ed.) *Environmental Design Research,* Vol. 2. Stroudsburg, Penna.: Dowden, Hutchinson & Ross, 1973.

Rapoport, A. Toward a redefinition of density. *Environment and Behavior,* 1975, *7,* 133–158.

Rapoport, A. Environmental cognition in cross-cultural perspective. In G.T. Moore & R.G. Golledge (Eds.) *Environmental Knowing: Theories, Research, and Methods.* Stroudsburg, Penna.: Dowden, Hutchinson & Ross, 1976.

Rapoport, A. *Human Aspects of Urban Form.* Oxford, England: Pergamon, 1977.

Rapoport, A. Cultural origins of architecture. In J.C. Snyder & A.J. Catanese (Eds.) *Introduction to Architecture.* New York: McGraw-Hill, 1979.

Rapoport, A. *The Meaning of the Built Environment: A Nonverbal Communication Approach.* Beverly Hills, Calif.: Sage, 1982.

Rapoport, A. & Watson, N.J. Cultural variability in physical standards. *Transactions of the Bartlett Society* (London), 1967–68, 63–83.

Reizenstein, J.E. Linking social research and design. *Journal of Architectural Research,* 1975, *4,* 26–38.

Reizenstein, J.E. & Zimring, C.M. (Eds.) Evaluating Occupied Environments. Special issue of *Environment and Behavior,* 1980, *12* (Whole No. 4).

Relph, E. *Place and Placelessness.* London: Pion, 1976.

Rice Center for Community Design and Research. *Environmental Design Research.* Houston: Rice Center, 1976.

Rittle, H.W.J. & Webber, M.M. Dilemmas in a general theory of planning. *Policy Sciences,* 1973, *4,* 155–169.

Rosaldo, M.Z. Women, culture, and society: a theoretical overview. In M.Z. Rosaldo and L. Lamphere (Eds.) *Women, Culture, and Society.* Stanford, Calif.: Stanford University Press, 1974.

Rosaldo, M.Z. & Lamphere, L. *Women, Culture, and Society.* Stanford, Calif.: Stanford University Press, 1974.

Ross, R.P. & Campbell, D.E. A review of the EDRA Proceedings. In W.E. Rogers & W.H. Ittelson (Eds.) *New Directions in Environment Design Research.* Washington, D.C: Environmental Design Research Association, 1978.

Rowles, G.D. *Prisoners of Space? Exploring the Geographical Experience of Older People.* Boulder, Colo.: Westview, 1978.

Russell, J.A. & Ward, L.M. Environmental psychology. *Annual Review of Psychology,* 1982, *33,* 651–688.

Rutledge, A.J. *Anatomy of a Park.* New York: McGraw-Hill, 1971.

Saarinen, T.F. *Environmental Planning: Perception and Behavior.* Boston: Houghton-Mifflin, 1976.

Saegert, S. & Winkel, G.H. The house: a critical problem for changing sex roles. In G.R. Wekerle, R. Peterson & D. Morley (Eds.) *New Space for Women.* Boulder, Colorado: Westview, 1980.

Saegert, S. Masculine cities and feminine suburbs: Polarized ideas, contradictory realities. In C.R. Stimpson, E. Dixler, M.J. Nelson & K.B. Yatrakis (Eds.) *Women and the American City.* Chicago: University of Chicago Press, 1980.

Salmen, J. *A Listing of Fifteen Suggested Priority Projects for Research in the Area of Barrier Free Design.* Final Report to the Community Services Administration, National Center for a Barrier Free Environment, 1980.

Sanoff, H. *Designing with Community Participation.* Stroudsburg, Penna.: Dowden, Hutchinson & Ross, 1978.

Schluntz, R. *Survey on Sponsored Architectural Research* (2 vols.). Washington, D.C.: Association of Collegiate Schools of Architecture, 1981.

Schorr, A.L. *Slums and Social Insecurity.* Washington, D.C: U.S. Government Printing Office, 1963.

Schwartz, B. (Ed.) *The Changing Face of Suburbs.* Chicago: University of Chicago Press, 1976.

Seamon, D. Body-subject, time-space routines, and place-ballets. In A. Buttimer & D. Seamon (Eds.) *The Human Experience of Space and Place.* London: Croom/Helm, 1980. (a)

Seamon D. *A Geography of the Lifeworld.* London: Croom/Helm, 1980. (b)

Seamon, D. & Nordin, C. Marketplace as place ballet. *Landscape,* 1980, *24*(3), 35–41.

Seidel, A.D. The credibility inherent in the use of various environment-behavior research techniques. In A.E. Osterberg, C.P. Tiernan, & R.A. Findlay (Eds.) *Design Research Interactions.* Washington, D.C: Environmental Design Research Association, 1981. (a)

Seidel, A.D. How designers evaluate information and why researchers may not care. In A.E. Osterberg, C.P. Tiernan & R.A. Findlay (Eds.) *Design Research Interactions.* Washington, D.C: Environmental Design Research Association, 1981. (b)

Seidel, A.D. Research is under-utilized because researchers and decision-makers have different conceptions of information quality. *Knowledge: Creation, Diffusion, Utilization,* 1981, *3,* 233–248. (c)

Seidel, A.D. Usuable EBR: What can we learn from other fields? In P. Bart, A. Chen, & G. Francescato (Eds.) *Knowledge for Design.* Washington, D.C: Environmental Design Research Association, 1982.

Simon, H.A. *The Sciences of the Artificial.* Cambridge, Mass.: MIT Press, 1969.

Simon, H.A. Style in design. In J. Archea & C.M. Eastman (Eds.) *Proceedings of the 2nd Annual Environmental Design Research Association Conference.* Stroudsburg, Penna.: Dowden, Hutchinson & Ross, 1970.

Simon, H.A. (Chair) *Social and Behavioral Science Programs in the National Science Foundation.* Washington, D.C: National Academy of Sciences, 1976.

Sims, B. Environmental design research: some emerging directions. *Man-Environment Systems,* March 1978, *8*(2), 67–74.

Skidmore, Owings & Merrill. *Program Options for Environmental Design Research.* Final Report to the National Science Foundation, Skidmore, Owings & Merrill, Boston, 1978.

Smith, M.B. Some problems of strategy in environmental psychology. In D. Stokols (Ed.) *Perspectives on Environment and Behavior.* New York: Plenum, 1977.

Snyder, J.C. (Ed.) *An Agenda for Architectural Research 1982.* Washington, D.C: Architectural Research Centers Consortium, n.d. (1982).

Snyder, J.C. (Ed.) *Architectural Research,* New York: Van Nostrand Reinhold, 1984.

Sommer, R. *Personal Space: The Behavioral Basis of Design.* Engelwood Cliffs, N.J.: Prentice-Hall, 1969.

Sommer, R. & Ross, H. Social interaction on a geriatrics ward. *International Journal of Social Psychiatry,* 1958, *4,* 128–133.

Stahl, F.I., Margulis, S.T., & Ventre, F.T. Ability of people to escape fires in public buildings: A review of code provisions and technical literature. Unpublished report, National Bureau of Standards, NBSIR 82-280, 1982.

Stea, D. Mediating the medium. *American Institute of Architects Journal,* 1967, *46,* 67–70.

Stea, D. The measurement of mental maps: An experimental model for studying conceptual spaces. In K.R. Cox & R.G. Golledge (Eds.) *Behavioral Problems in Geography.* Evanston, Ill.: Northwestern University Studies in Geography No. 17, 1969.

Stea, D. The use of environmental modelling ("toy play") for studying the environmental cognition of children and adults. In *Proceedings of the 14th Congress,* Sociedad Interamericana de Psicologia, São Paulo, Brasil, 1975.

Stea, D. Environmental cognition and the architecture of human settlement: An interdisciplinary and cross-cultural approach. In W.H. Rogers & W.H. Ittelson (Eds.) *New Directions in Environmental Design*

Research. Washington, D.C: Environmental Design Research Association, 1978.

Steinfeld, E. *Access to the Built Environment* (7 vols.). Washington, D.C: U.S. Government Printing Office, and U.S. Department of Housing and Urban Development, 1979.

Stimpson, C.R., Dixler, E., Nelson, M.J., & Yatrakis, K.B. (Eds.). *Women and the American City*. Chicago: University of Chicago Press, 1980.

Stokols, D. On the distinction between density and crowding: Some implications for future research. *Psychological Review*, 1972, *79*, 275–278. (a)

Stokols, D. A social psychological model of human crowding phenomena. *Journal of the American Institute of Planners*, 1972, *38*, 72–83. (b)

Stokols, D. Toward an operational theory on alienation. *Psychological Review*, 1975, *82*.

Stokols, D. Origins and directions of environment-behavioral research. In D. Stokols (Ed.) *Perspectives on Environment and Behavior: Theory, Research and Applications*. New York: Plenum, 1977. (a)

Stokols, D. (Ed.) *Perspectives on Environment and Behavior*. New York: Plenum Press, 1977. (b)

Stokols, D. Environmental psychology. *Annual Review of Psychology*, 1978, *29*, 253–296.

Stokols, D. A congruence analysis of human stress. In I.G. Sarason & C.D. Speilberger (Eds.) *Stress and Anxiety*. Washington, D.C: Hemisphere Publishing, 1979.

Stokols, D. Environmental psychology: A coming of age. M.A. Kraut (Ed.) *G. Stanley Hall Lecture Series*, Vol. 2. Washington, D.C: American Psychological Association, 1981. (a)

Stokols, D. Group x place transactions: Some neglected issues in psychological research on settings. In D. Magnussen (Ed.) *Toward a Psychology of Situations: An Interactional Perspective*. Hillsdale, N.J.: Erlbaum, 1981. (b)

Stokols, D. & Shumaker, S.H. People in places: A transactional view of settings. In J.H. Harvey (Ed.) *Cognition, Social Behavior, and the Environment*. Hillsdale, N.J.: Erlbaum, 1981.

Studer, R.G. & Stea, D. *Directory of Behavior and Environmental Design*. Providence, R.I.: Brown University, 1965.

Sundstrom, E., Kastenbaum, D. & Konar-Goldband, E. *Physical Office Environments, Employee Satisfaction, and Job Performance*. Final Report to the American Institute of Architects Research Corporation, Buffalo Organization for Social and Technological Innovation, Buffalo, N.Y., 1978.

Szigeti, F. & Davis, G. A conceptual framework for the knowledge base relevant to environmental design research. Paper presented at the Environmental Design Research Association Conference, Charleston, South Carolina, April 1980.

Toffler, A. *The Third Wave*. New York: Morrow, 1980.

Tuan, Y.-F. *Topophilia: A Study of Environmental Perception, Attitudes, and Values*. Englewood Cliffs, N.J.: Prentice-Hall, 1974.

Tuan, Y.-F. *Space and Place: The Perspective of Experience.* Minneapolis: University of Minnesota Press, 1977.

Tuttle, D.P. Suburban Fantasies. Unpublished M.Arch Thesis, Department of Architecture, University of Wisconsin-Milwaukee, 1983.

U.S. Bureau of the Census. *Statistical Abstracts of the United States.* Washington, D.C.: U.S. Government Printing Office, 1980.

U.S. Consumer Product Safety Commission. *Hazard Analysis of Injuries Relating to Playground Equipment.* Washington, D.C.: Author, 1975.

U.S. Consumer Product Safety Commission. *A Handbook for Public Playground Safety* (2 vols.). Washington, D.C.: U.S. Government Printing Office, 1981.

Ventre, F.T. Transforming environmental research into regulatory policy. In B. Honikman (Ed.) *Responding to Social Change.* Stroudsburg, Penna.: Dowden, Hutchinson & Ross, 1975.

Ventre, F.T. Building in eclipse: Architecture in secession. *Progressive Architecture,* 1982, *62,* 58–61.

Ventre, F.T., Stahl, F.I. & Turner, G.E. Crowd ingress to places of assembly: Summary and proceedings of an experts' workshop. Unpublished report, National Bureau of Standards, NBSIR 81-2361, 1982.

Villecco, M. & Brill, M. *Environmental Design/Research: Concepts, Methods, and Values.* Washington, D.C.: National Endowment for the Arts, 1981.

Wapner, S. Transactions of persons in environments: some critical transitions. *Journal of Environmental Psychology,* 1981, *1,* 223–239.

Wapner, S., Cohen, S.B. & Kaplan, B. (Eds.) *Experiencing the Environment.* New York: Plenum, 1976.

Wapner, S., Kaplan, B. & Cohen, S.B. Exploratory applications of the organismic-developmental approach to the transactions of men-in-environments. In S. Wapner, S.B. Cohen & B. Kaplan (Eds.) *Experiencing the Environment.* New York: Plenum, 1976.

Warfield, J. *Designs for the Future of Environmental Education,* Vol. 1. Final report to the Office of Environmental Education, U.S. Department of Education, Washington, D.C., July 1980.

Webb, E.J., Campbell, D.T., Schwartz, R.D. & Sechrest, L. *Unobtrusive Measures: Nonreactive Research in the Social Sciences.* Chicago: Rand McNally, 1966.

Webber, M.M., Dyckman, J.W., Foley, D.L., Guttenberg, A.Z., Wheaton, W.L.C. & Bauer Wurster, C. (Eds.). *Explorations into Urban Structure.* Philadelphia: University of Pennsylvania Press, 1964.

Weidemann, S., Anderson, J.R., Butterfield, D.I. & O'Donnell, P.M. Residents' perception of satisfaction and safety: A basis for change in multifamily housing. In G.T. Moore (Ed.) Applied Architectural Research: Post-Occupancy Evaluation of Buildings. Special issue of *Environment and Behavior,* 1982, *14,* 615–724.

Weinstein, C.S. The physical environment of the school: A review of the research. *Review of Educational Research,* 1979, *49,* 577–610.

Weinstein, C.S. & David, T.G. (Eds.) *Spaces for Children: The Built Environment and Child Development.* New York: Plenum, in press.

Weiss, C.H. (Ed.) *Using Social Research in Public Policy Making*. Lexington, Mass.: Lexington 1977.

Weiss, C.H. & Barton, A.H. (Eds.) *Making Bureaucracies Work*. Beverly Hills, Calif.: Sage, 1979.

Wekerle, R., Peterson, R. & Morley, D. (Eds.) *New Space for Women*. Boulder, Colo.: Westview, 1980.

Wener, R.E. Post-occupancy evaluation success stories. Unpublished manuscript, Department of Psychology, Polytechnic Institute of New York, 1982.

White, W.P. (Ed.) *Resources in Environment and Behavior*. Washington, D.C.: American Psychological Association, 1979.

White House Conference on Aging. *Chartbook on Aging in America*. Washington, D.C.: U.S. White House, 1981.

Whyte, W.H. *The Social Life of Small Urban Spaces*. Washington, D.C.: Conservation Foundation, 1980.

Wicker, A.W. *An Introduction to Ecological Psychology*. Monterey, Calif.: Brooks/Cole, 1979.

Wicker, A.W. & Kirmeyer, S.L. From church to laboratory to national park. In S. Wapner, S.B. Cohen & B. Kaplan (Eds.) *Experiencing the Environment*. New York: Plenum, 1976.

Wildavsky, A.B. If planning is everything, maybe it's nothing. *Policy Sciences*, 1973, *4*, 127–153.

Wildavsky, A.B. *The Politics of the Budgetary Process*. Boston: Little Brown, 1974.

Wilensky, H.L. *Organizational Intelligence: Knowledge and Policy in Government and Industry*. New York: Basic Books, 1976.

Willems, E.P. Behavioral ecology. In D. Stokols (Ed.) *Perspectives on Environment and Behavior: Theories, Research, and Applications*. New York: Plenum, 1977.

Wineman, J.D. (Ed.) Office Design and Evaluation. Special issues of *Environment and Behavior*, 1982, *14* (Whole Nos. 3 and 5).

Wohlwill, J.F. The emerging discipline of environmental psychology. *American Psychologist*, 1970, *25*, 303–312.

Wohlwill, J.F. & Kohn, I. Dimensionalizing the environmental manifold. In S. Wapner, B. Kaplan, & S.B. Cohen (Eds.) *Experiencing the Environment*. New York: Plenum, 1976.

Wright, J.K. Terrae incognitae: The place of imagination in geography. *Annuals of the Association of American Geographers*, 1947, *37*, 1–15.

Zeisel, J. Fundamental values in planning with the nonpaying client. In J. Lang, C. Burnette, W. Moleski & D. Vachon (Eds.). *Designing for Human Behavior: Architecture and the Behavioral Sciences*. Stroudsburg, Penna.: Dowden, Hutchinson & Ross, 1974.

Zeisel, J. *Sociology and Architectural Design*. New York: Russell Sage Foundation, 1975.

Zeisel, J. *Stopping School Property Damage: Design and Administrative Guidelines to Reduce School Vandalism*. Arlington, Va.: American Association of School Administrators, 1976.

Zeisel, J. *Inquiry By Design: Tools for Environment-Behavior Research.* Monterey, Calif.: Brooks/Cole, 1981.

Zeisel, J., Epp., G. & Demos, S. *Low-Rise Housing for Older People.* Washington, D.C.: U.S. Government Printing Office and U.S. Department of Housing and Urban Development, 1978.

Zeisel, J., Epp, G. & Demos, S. *Mid-Rise Elevator Housing for Older People.* Washington, D.C., U.S. Government Printing Office and U.S. Department of Housing and Urban Development, 1984.

Zeisel, J. & Griffin, M. *Charlesview Housing: A Diagnostic Evaluation.* Cambridge, Mass.: Harvard University, Architectural Research Office, 1975.

Zube, E.H. (Ed.) *Landscapes: Selected Writings of J.B. Jackson.* Amherst, Mass.: University of Massachusetts Press, 1970.

Zube, E.H. Perception of landscape and land use. In I. Altman & J.F. Wohlwill (Eds.) *Human Behavior and Environment: Advances in Theory and Research.* New York: Plenum, 1976.

Zube, E.H. *Environmental Evaluation: Perception and Public Policy.* Monterey, Calif.: Brooks/Cole, 1980. (a)

Zube, E.H. The roots of future innovations: Research and theory. *Landscape Architecture,* 1980, *70,* 614–617. (b)

Zube, E.H. (Ed.) *Social Sciences, Interdisciplinary Research, and the U.S. Man and the Biosphere Program.* Washington, D.C.: U.S. Department of State, Man and the Biosphere Secretariat, 1980. (c)

Zube, E.H., Brush, R.O. & Fabos, J.G. (Eds.) *Landscape Assessment: Value, Perceptions, and Resources.* Stroudsburg, Penna.: Dowden, Hutchinson & Ross, 1975.

Index

About the Authors

Gary T. Moore is a research architect and environmental psychologist. A cofounder of the Environmental Design Research Association (EDRA) in 1968, he was Co-Chair of the Steering Committee from 1968 to 1970, has held several other offices since then, was Chair of the Board of Directors in 1980–81, and is now Co-Managing Editor of two EDRA publication series. He holds degrees in architecture (B.Arch., University of California, Berkeley) and environmental psychology (M.A., Ph.D., Clark University) and has held academic positions at Clark University, the University of Oregon, at Sydney University and the University of New South Wales in Australia, and at Victoria University of Wellington in New Zealand. Currently he is Associate Professor of Architecture, Coordinator of the Ph.D. Program, and Director of the Center for Architecture and Urban Planning Research at the University of Wisconsin-Milwaukee. He is the author of three previous books and many chapters and papers on environment, behavior, and design. His research focuses on alternative housing for the elderly and on life-span development and the environment, including the preparation of a book on *Children and the Physical Environment* and a series of papers on environmental gerontology. He is co-editing a series of books, "Advances in Environment, Behavior, and Design," is an Associate Editor of *Environment and Behavior*, and is an Advisory Editor for Praeger Publishers.

D. Paul Tuttle is an urban planner and architect. He holds professional degrees in urban planning (B.S. in City and Regional Planning, California Polytechnic State University; M.U.P., San Jose State University) and in architecture (M.Arch., University of Wisconsin-Milwaukee). His professional work has been in architecture, urban design, and urban planning, mostly in California, including being Associate Planning Director for the cities of Pinole and Vallejo, California. His research focuses on housing and urban development and on ordinary people and their environments. Currently he is a Ph.D. student in environment and behavior at the University of California, Berkeley.

Sandra C. Howell is an environmental psychologist and consultant in design, planning, and research. She holds degrees in public health (M.P.H., University of California, Berkeley) and psychology (Ph.D., Washington University, St. Louis). She has been active in research and organizational activities on health and the environment, urban housing, and aging and the environment for many years, including being active in EDRA as a member of the Board of Directors and in the Gerontological Society as an officer of the Section on Behavioral and Social Sciences. At present she is Associate Professor of Behavioral Sciences in Architecture at the Massachusetts Institute of Technology. She is the author of two previous books and many chapters and papers. Currently her research focuses on the evaluation of and program preparation for the development of housing for the severely and profoundly retarded, and she is preparing a book called *Habit and Habitability* based on cross-cultural and life-cycle analysis of domestic events.